TAKING RELIGION SERIOUSLY

TAKING RELIGION SERIOUSLY

CHARLES MURRAY

Encounter BOOKS

New York · London

Copyright © 2025 by Cox and Murray, Inc.

All rights reserved. No part of this publication may be reproduced, stored in a retrieval system, or transmitted, in any form or by any means, electronic, mechanical, photocopying, recording, or otherwise, without the prior written permission of Encounter Books, 900 Broadway, Suite 601, New York, New York, 10003.

First American edition published in 2025 by Encounter Books, an activity of Encounter for Culture and Education, Inc., a nonprofit, tax-exempt corporation. Encounter Books website address: www.encounterbooks.com

Manufactured in the United States and printed on acid-free paper. The paper used in this publication meets the minimum requirements of ANSI/NISO Z39.48–1992 (R 1997) (*Permanence of Paper*).

First American Edition

Library of Congress cataloging-in-publication data is available

Information for this title can be found at the Library of Congress website under the following ISBN 978-1-64177-485-7 and LCCN 2025029720.

Table of Contents

	Introduction	1

Part I: Taking God Seriously

1	"More Than Evolution Required"	7
2	Perceptual Deficit	13
3	Moving off Dead Center	21
4	The Brute Facts of the Big Bang	29
5	Challenges to Materialism	45

Part II: Taking Christianity Seriously

6	A Strange New Respect	63
7	Enter C. S. Lewis	73
8	The Moral Law	77
9	Who Wrote the Gospels and When?	85
10	The Historicity of the Gospels	111
11	What's the Point?	141
	Acknowledgments	159
	Notes	161
	Index	179

To Arthur Brooks

Introduction

"It ought to be your next book," Nick said.

Nicholas Eberstadt and Karlyn Bowman, friends and colleagues for more than three decades, had just finished interviewing me about my years at the American Enterprise Institute. Late in the interview, we had somehow gotten onto religion. I had described a little about my evolution from agnosticism to Christianity. Nick, a Roman Catholic, found my eccentric and haphazard collection of ideas to be entertaining.

Nick's suggestion was immediately attractive. At the time, I was working on an autobiographical book about my role in the conservative and libertarian intellectual movements of the 1980s and 1990s. I was about to give it up. My career has had memorable moments, but not a book's worth. I couldn't imagine that many people would be interested. In contrast, millions are like me when it comes to religion: well-educated and successful people for whom religion has been irrelevant. We grew up in secular households or drifted away from the faiths in which we were raised and never looked back. For them, I think I have a story worth telling.

In my case, I attended Presbyterian Sunday School and, later, church services into adolescence but then went off to Harvard, where I was as thoroughly socialized to be secular as earlier generations of Harvard students had been socialized to be devout. By *socialized*, I don't mean that my professors tried to convince me that Thomas Aquinas was wrong. I didn't study religion at all. None of the professors I admired was religious (at least visibly). I didn't have a single friend who was religious. When the topic of religion came up, professors and friends alike treated it dismissively or as a subject for humor. I fit into the *zeitgeist*.

If asked, I would have said I was an agnostic, but I didn't spend much time thinking about religion because I couldn't see the point. If God exists, he could not be the kind of God who has anything to do with this flyspeck world, let alone with the lives of the individual human beings clinging to its surface.

I've changed my mind about God in general (that's the subject of Part I of this book) and about Christianity specifically (Part II), but I don't proselytize. Rather, I urge upon you that religion is something that should be taken seriously by nonbelievers and that *can* be taken seriously in the same way that Chinese history or plate tectonics can be taken seriously—by reading a lot, thinking about what you've read, and bouncing your reactions off people who know more than you do. I am about to describe how it worked for me in hopes that my experience may be useful to you. If nothing else, you're going to get some good reading lists.

Introduction

    My meandering pilgrimage to belief has been devoid of divine revelation. Many people assume, as I once did, that to start taking religion seriously requires some sort of born-again moment. Not necessarily. I am reminded of a long-ago conversation with the late Catholic social philosopher Michael Novak. I had expressed my dismay at Catholicism's insistence on sticking with unbelievable doctrines like transubstantiation. Michael said, "God needs a church that can speak to everyone." Maybe God also needs a way to reach over-educated agnostics, and that's what I stumbled into. It's a more arid process than divine revelation but it has been rewarding. And, if you're like me, it's the only game in town.

Charles Murray
Burkittsville, Maryland
27 April 2025

PART I

TAKING GOD SERIOUSLY

CHAPTER 1

"More Than Evolution Required"

When our daughter Anna was born in July 1985, I was two years into a marriage that had made me happier than I had ever been and nine months into the unexpected success of *Losing Ground*. My two daughters from a previous marriage loved Catherine, their new stepmother, and were enthusiastic about having a new sister. I had no sense of anything lacking in my life, least of all religion. I was a happy agnostic.

It was Catherine who had the epiphany that set things in motion. During the first few months after Anna's birth, Catherine discovered that her love for her daughter surpassed anything she had ever known. It was a love so all-enveloping, she later told me, that she had trouble distinguishing where she stopped and Anna began. Her epiphany was that she loved Anna "far more than evolution required."

That's the way she put it. It was a great line, picked up and quoted years later by both David Brooks and Michael Gerson. In saying it, she acknowledged that evolution dictates that women who love their infants are more likely to pass on their genes than women who don't, but she also felt as if she were a conduit for a greater love. That greater love, she decided, pointed vaguely toward God.

Catherine's religious history up to that point paralleled mine. She grew up in Newton, Iowa, as had I (I'm six years older), and attended the Methodist Church located next door to my family's Presbyterian Church. She was fervently religious into her high school years, but then she left Newton for the University of Iowa where, again in her words, "I learned that smart people don't believe that stuff anymore." T. H. Huxley, the nineteenth-century British scientist who coined the word *agnostic*, became her intellectual hero.

Catherine had also become angry at the hypocrisy of organized Christianity. After her epiphany, this was a problem. She wanted to explore the tenuous hypothesis that was taking shape in her mind, and she wanted a structure within which she could do that, but she was allergic to doctrine. She tried visiting the Sunday services of some of the mainline Christian denominations near our downtown Washington neighborhood. They didn't work for what she needed. Among other things, she couldn't in good conscience recite the Apostles' Creed along with the rest of the congregation.

Then one Sunday she visited the Friends Meeting of Washington, located on Florida Avenue in downtown DC, and experienced her first Quaker meeting. It was completely unlike any religious service she had ever attended. People were seated on benches forming a square. Occasionally someone rose and offered a brief reflection or read a short quotation or prayer. That was it. No Apostles' Creed. Also no pastor, no choir, no hymns, no sermon, no benediction.

Theologically, Quakerism is somewhere between Unitarianism and mainline Protestantism. It is not even required that members of the meeting believe in God, but a Quaker meeting is likely to have several members who subscribe to many elements of Christian theology, and Friends are drawn to Christian moral teachings. Quakerism offered Catherine the kind of nondoctrinaire spiritual environment she was looking for.

EDWARD BANFIELD ON UNITARIANISM

In the late 1980s, I was on a panel with two eminent academics, Allan Bloom and Edward Banfield. During the Q&A, someone wanted to know our positions on God. Bloom, caught off guard, mumbled something incoherent. Banfield was unfazed. "I was raised as a Unitarian," he said. "Unitarians are taught to believe in one God . . . at most." In the laughter that followed, the moderator forgot to ask me. Thank God.

Catherine attended the Florida Avenue Meeting for several months. Then she started driving out to Bethesda Friends' Meeting in suburban Maryland, a quieter place. I stayed home on Sunday—"First Day" to Quakers—and looked after Anna. I joined Catherine for a meeting only once, just before Christmas in 1988. In my daily log, the entry concluded with "Not sure what I think. Could be very nice. Could be grating."

Taking Religion Seriously

In November 1989, Catherine, four-year-old Anna, our infant son Bennett, and I moved from our townhouse in Adams Morgan, a funky urban neighborhood, to a village of about 150 people an hour's drive northwest of Washington's Beltway. We weren't fleeing the big city for an arcadian refuge. We were broke.

The salary schedules for scholars at think tanks are like academia's, though without the tenure. My salary in 1989 was closer to an assistant professor's than a full professor's. We had hopes that a journalistic account of the Apollo lunar program we were writing together would bring in some big royalty checks. The book, *Apollo: The Race to the Moon*, was published in July 1989, to gratifyingly rave reviews. But it didn't sell. We couldn't afford to raise our children in Adams Morgan. We would have to move.

Neither of us wanted to live in a suburb, so we went searching for something else. We found Burkittsville, a nineteenth-century village located at the foot of South Mountain in Frederick County. We bought a house on the edge of town with four acres and a pond for considerably less than we got for our townhouse with no parking space in Adams Morgan. Thirty-six years later, we're still there.

Catherine went looking for a new Quaker meeting. She tried the one in Frederick, the closest big town, twenty miles away. It was welcoming, but she was still restless. Then she visited another meeting she had heard about, Goose Creek Meeting in Lincoln, Virginia. The first Goose Creek Meeting House was built of logs in the mid-eighteenth century. As far as anyone can tell, Goose Creek Meeting has conducted a meeting for worship every First Day since then, in the same building since 1819. The meeting is named after the village of Goose Creek, which changed its name to Lincoln just after Abraham Lincoln's election in 1860. That a Virginia town would do that in the fall of 1860 tells you all you need to know about how devoutly Quaker its inhabitants were.

Many of the members of Goose Creek Meeting were direct descendants of the meeting's founders. Catherine loved that sense of rootedness. She was conscious of being a cliché—a secular Baby Boomer reaching middle age seeking a spiritual life—but she was welcomed by Friends whose Quaker forebears had been sitting on the same benches for generations.

Catherine started driving the twenty-two miles from Burkittsville to Goose Creek Meeting in the spring of 1990, taking Anna with her to First Day School. I seldom went along, instead looking after our new son and working on a book about IQ that I was writing in collaboration with Richard Herrnstein. I was still sure that Catherine's growing interest in things of the spirit had nothing to do with me.

CHAPTER 2

Perceptual Deficit

By the beginning of October 1993, I had no more excuses to spend First Days at home. Dick Herrnstein and I had completed the draft of our book, which now had a title: *The Bell Curve*. It was time for me to participate in this important part of my wife and children's life, godless though I still was. My log entry for Sunday, October 3, began, "To meeting, now and forever." I made good on it, mostly, until both of our children had left home for college.

The origin of my decision to start attending Meeting traced back to a conversation with my father thirty years earlier on a visit home during my sophomore year in college. Dad had been at the wheel, driving my mother and me to church. I announced that while I was happy to attend services with them when I was in Newton, they should know that I no longer believed in Christianity. I thought I was being daringly candid and grown-up. The word *sophomoric* exists for a reason.

Dad replied that he didn't either—"I don't believe I'm going to go to heaven when I die or any of the rest of it." Why then had he been going to church every week year after year? He replied that it was a good habit, a kind of discipline. I cannot remember his exact words, but the gist was that attending church reminded him of ways he could be a better person and fulfill his responsibilities to others (as if he had ever shirked a responsibility in his life).

His words were still in my mind in October 1993, along with my belief that growing up within a religious tradition was a good thing for children. But I took my resolve to be an observant Quaker without much optimism that it would do me any good. I had discovered when I was still in my twenties that I don't have much talent for spiritual contemplation, and that realization had only been confirmed by my sporadic attendance at Quaker meetings.

I graduated from college in early June 1965 and flew to Hilo, Hawaii, for Peace Corps training the day after commencement. I left Hilo for my assignment with the Royal Thai Ministry of Health's Village Health and Sanitation Project in September. Except for a two-week visit home in 1968, I didn't return to the States until August 1970. In effect, I missed the years that Americans have in mind when they talk about "the Sixties."

Over the course of those five years in Thailand, I got caught up in my generation's attraction to transcendental meditation. Thailand didn't have celebrity

Perceptual Deficit

meditators like the Beatles' favorite, Maharishi Mahesh Yogi, but Thai Buddhist monks routinely practiced classic meditation techniques.

I was especially attracted by the way Buddhism didn't try to convince you of its doctrines. As I understood it, Buddhism simply said that if you quiet your mind and become able to perceive reality, you will find that reality corresponds to Buddhist teachings. You don't need to take anything on faith.

I was too shy to go to a temple and seek out a teacher. Instead, I bought some books on the subject and started practicing meditation as best I could. At some point, a mutual friend introduced me to Peter the Monk. Peter the Monk was an American in his early thirties who had roamed Asia for a few years and ended up as a monk in a Thai temple. He was the real thing, not a glorified backpacker. He had already been in the temple for two years and was adept at reaching deep meditative states. He tried to tutor me. It didn't work. Occasionally I would get close to something that felt like a meditative state, but no further.

I did have one experience that taught me something I had suspected but not known for sure. I liked music, but superficially. I was seldom moved by music. During performances of long classical pieces I usually drifted off into my own thoughts. One night I put a recording of one of Beethoven's symphonies on the phonograph and resolved to treat it as meditation, staying in the moment from beginning to end.

I did, and it was like no other listening experience

in my life. At the end, I said to myself, "This must be how some people always hear music." It never happened again, but it got me thinking about something that has expanded into a semi-coherent theory over the years: Just as people have different levels of cognitive ability or athletic coordination, so too they have different levels of perceptual ability specific to many different domains.

In the appreciation of music, the top level consists of the ability of to enter wholly into the music, to fully appreciate what is going on with harmony, rhythm, and the interplay of instruments and voices. The bottom level is tone deafness. The middle level is represented by the kind of casual enjoyment that I ordinarily got from music.

The same is true of appreciating the visual arts and great literature. I'm not talking about IQ. People with stratospheric IQs can be tone deaf, unmoved by great art, and think Shakespeare is boring.[1] The only similarity between cognitive ability and various perceptual abilities is that they come in gradations from low to high. So does spirituality. Some people can perceive spiritual realities that others can't.

Most atheists I have known reject that proposition. They are certain that people who hold deep religious beliefs are deluding themselves. Being married to Catherine, I didn't have that option. She had an extraordinary intellect, was fully self-aware, and wasn't deluding herself in any way. Through her own example and the example of people I got to know through her, I had come to accept that I was the one with a problem. I suffer from a perceptual deficit in spirituality.

Part of the explanation is an extreme form of individualism that has been part of my makeup since my earliest memories. It surfaced with a vengeance in my futile attempts to meditate. On those rare occasions when I came close to a true meditative state, I could feel myself resisting. The idea of giving up that much of my autonomy scared me, however much I could argue with myself that meditation augments mental autonomy.

A similar thing happened during the nineties when I once made a concerted effort to pray. I can't recall what problem prompted it, but I do remember that I prayed earnestly and at length. In the days that followed I felt at peace with whatever it was that had been worrying me. I was confident of having resolved it. I never tried to pray again. It was scary—not because prayer had failed, but because it had succeeded.

When Catherine read an early draft of this chapter, she observed that she likes being in control as much as I do (which indeed she does). The difference between us, she said, was that her sense of need for belief was greater. I agreed with that, and I also had a suspicion about why. I had distracted myself with Western modernity.

I am using *Western modernity* as shorthand for all the ways in which life in the last hundred years has shielded many of us from the agonizing losses, pains, and sorrows that were part of human life since the dawn of humankind. Parents before the twentieth century commonly suffered the death of a child. Husbands commonly lost beloved wives to childbirth, and wives commonly lost beloved husbands to work accidents or war. Everyone lived their lives from adolescence onward knowing that waking up with a sore throat could mean that they would die before the week was out. Injuries that now mean surgery and a few days in the hospital meant permanent disability and often permanent pain.

Most people still suffer at least one such agonizing event eventually, but often not until old age and sometimes never. So far, that's been the case with me. I've lived my life without ever reaching the depths of despair. I'm grateful for my luck. But I have also not felt the God-sized hole in my life that the depths of despair often reveal. This doesn't mean there isn't a hole; just that I've been able to ignore it. And that, I think, explains a lot not only about me but about the unreflective secularism of our age.

Catherine did not require despair to recognize her God-sized hole, and she did not ignore it. As the nineties went on and I watched her progress on her spiritual journey, I realized that I couldn't keep up with her. I didn't get it.

At the same time, I yearned to participate. I couldn't do it in the same way, but I could do it more circuitously. I began to think in terms of an analogy with the way I deal with my limits in mathematics.

My math ability is okay, but not good enough to allow

me to understand the nuances of statistical probability just by studying the equations. I need concrete examples. But given those concrete examples, I have found that I can deploy my other mental assets to understand what the math is doing. Similarly, I cannot enter into a journey involving religious faith by the path that people who are more receptive to spiritual experiences can use. But I can deploy alternative strategies. That's what I've done and that's what I'm about to describe.

CHAPTER 3

Moving off Dead Center

My secular catechism from college through the mid-1990s went something like this:

> The concept of a personal God is at odds with everything that science has taught us over the last five centuries. Earth is a nondescript planet on the edge of a nondescript galaxy in a universe with billions of galaxies.
>
> Humans are animals, different from other animals in our level of consciousness and intelligence but evolved from the same primitive life forms and subject to the same laws of biology and physics. Our thoughts and emotions are produced by the brain. When the brain stops, consciousness stops too.
>
> The great religious traditions are human inventions, natural products of the fear of death. That includes Christianity, which can call on no solid evidence for its implausible claims. The moral teachings attributed to Jesus are admirable, but many of them resemble other influential teachings drawn from Western and Eastern traditions.

I look back on that catechism and call it *dead center* because it was so unreflective. I had not investigated the factual validity any of those propositions. They were part of the received wisdom of most Western intellectuals throughout the twentieth century. I accepted them without thinking.

In describing how I got unstuck, I will make the process sound more orderly than it was. The actual process was a series of doubts about my settled answers that bubbled up periodically throughout the last half of the 1990s and the early 2000s. I experienced a series of nudges spread over many years, and they do not form a coherent whole.

Mathematics and Reality

The first nudge, so soft that it barely registered (I cannot recall when it did more than cross my mind) was the mathematical simplicity of many scientific phenomena—most famously $E = mc^2$, but also $F = ma$ (Newton's Second Law of Motion), $d = \frac{1}{2}gt^2$ (Galileo's Law), $V = IR$ (Ohm's Law), $P_1V_1 = P_2V_2$ (Boyle's Law), and a few dozen others.

I didn't realize it at the time, but the relationship of mathematics to natural phenomena has many deeper layers. For example, a Hilbert space was defined by pure mathematics that David Hilbert and others developed without reference to any practical applications—and it later turned out to be an indispensable tool for Paul Dirac's concepts of quantum mechanics. The relationship of mathematics to reality has generated shelves of books of which I was entirely ignorant (and remain nearly so). But even at my elementary level, it just seemed extremely odd that so many basic phenomena were so mathematically simple. It was almost as if someone had planned it that way.

SOURCES FOR THE MATHEMATICS OF REALITY

The nudge wasn't strong enough to prompt any reading at the time. Three sources that can get you started are a famous essay by Eugene Wigner, "The Unreasonable Effectiveness of Mathematics in the Natural Sciences" (1960),[1] Ian Hacking's *Why Is There Philosophy of Mathematics at All?* (2014), and Graham Farmelo's *The Universe Speaks in Numbers: How Modern Math Reveals Nature's Deepest Secrets* (2019).

"WHY IS THERE SOMETHING RATHER THAN NOTHING?"

The first unmistakable nudge involved the question, "Why is there something rather than nothing?" I first heard it put in those words by the late columnist and commentator Charles Krauthammer during a session of the Pariah Chess Club, probably around 1994 or 1995.[2] That I thought Charles had come up with it himself is proof of how unreflective I had been. Anyone who had taken any interest in theology would have encountered it long since. It's one of the most famous questions in metaphysics.[3]

But I hadn't heard it, and it caught me by surprise. When I had thought about the existence of the universe at all, I had taken it as a given. I am alive, I am surrounded by the world, the fact that I can ask the question presupposes that the universe exists. There's nothing else to be said. It is a mystery with a lower-case *m*.

Hearing the question stated so baldly and so eloquently made me start to take the issue seriously. Why is there *anything?* Surely things do not exist without having been created. What created all this? If you haven't thought about it recently, this is a good time to stop and try to come up with your own answer.

Whatever that answer may be, it is vulnerable to an infinite regress. What created the force behind the creation? Even if your answer is "God," you must ask how God came to be. At that point you're stuck with saying that it's turtles all the way down.

"It's turtles all the way down" is the punchline for a joke with variations that go back centuries. Modern versions of it begin with a distinguished scientist giving a lecture on astronomy. Here's how Stephen Hawking told it in *A Brief History of Time*:

> At the end of the lecture, a little old lady at the back of the room got up and said: "What you have told us is rubbish. The world is really a flat plate supported on the back of a giant tortoise." The scientist gave a superior smile before replying, "What is the tortoise standing on?" "You're very clever, young man, very clever," said the old lady. "But it's turtles all the way down!"[4]

Some eminent thinkers have argued that the question about existence is meaningless; others, that the universe did not require an act of creation. The box recommends two contemporary discussions. Both were published after I had already reached two conclusions that reading them did not change.

SOURCES FOR SECULAR ANSWERS TO "WHY IS THERE SOMETHING...?"

Jim Holt's *Why Does the World Exist? An Existential Detective Story* (2012) is an engagingly written and wide-ranging exploration of the alternative answers. For a more formal presentation, see Lawrence M. Krauss's *A Universe from Nothing: Why There Is Something Rather than Nothing* (2012).

One conclusion was that the existence of anything rather than nothing is a Mystery with a capital *M*. The second was that there is no meaningful difference between using the words *Mystery* and *God*. Saying that God created the universe does not tie me into any theological position. Using the word *God* amounts to truth in labeling in a way that *Mystery* does not. *Mystery* is a weasel word. I haven't any good explanation for what could have caused the universe, but I believe there must have been a cause, and I recognize that any answer the human brain can comprehend runs into the turtles-all-the-way-down problem. What *Mystery* really means is that the universe was created by an unknowable creative force that itself has no explainable source. Aristotle's unmoved mover. By the late 1990s, that sounded to me like a description of God I could accept.

DE-ANTHROPOMORPHIZING GOD

My ruminations about "Why is there something rather than nothing?" had a side effect. They helped me to stop anthropomorphizing God and instead give him the respect he deserves.

The Bible relentlessly anthropomorphizes God, starting in Genesis with the assertion that God created man in his own image. The God of the Old Testament has the full range of human characteristics—he gets angry, changes his mind, is remorseful, commands people to take vengeance on enemies, and tests the faith of Abraham and Job in ways that look a lot like cruelty.[5]

The New Testament's verbal imagery of God as a father and Jesus sitting at God's right hand reinforce the anthropomorphic view of God. That image has been reinforced still further by Christian art—think of Michelangelo's Sistine Chapel depiction of God as a formidable old man with flowing hair touching Adam's finger.

None of that had ever made sense to me. Once I decided that there had to be an unmoved mover and was intellectually committed to accepting that conception of God, I was free to think about a truth that, once you stop to think about it, *must* be a truth: Any God worthy of the name is at least as incomprehensible to a human being as I am to my dog. The analogy is better than it may seem at first glance. My dog is smart enough to perceive a few things about me—the fact that I exist as a distinct individual and that I feed her every morning. She also has some perceptions about my moods and what I want her to do. But these understandings represent only a few trivial aspects of who I am. I am not invisible to my dog, just as God is not invisible to me (I have come to believe), but I am nonetheless unknowable to my dog in any meaningful sense. God is just as unknowable to me.

Two other useful concepts entered my thinking sometime during the 1990s. One was that God exists outside of time—as taught by Aristotle but elaborated by Thomas Aquinas (who also taught that time began with the creation of the universe—something that is accepted by today's physicists). Just trying to get your head around the concept of existing outside time is a good way to realize how unknowable a being we are talking about.

Quaker teachings are also helpful in de-anthropomorphizing God. They emphasize that God is not a being with a location. He is everywhere—not just watching from everywhere but permeating the universe and our world. And there is the most famous of Quaker precepts: "There is that of God in everyone." It is not the same as saying, "There's some good in everyone." God is *in* you in some sense, along with permeating everything else. These are not concepts that can be fully processed (at least by me), but they are powerful antidotes to thinking about God as an especially wise and powerful grandpa.

Such were my cautious, tip-toeing steps into the

shallow end of the pool as of the early 2000s. I was about to find myself in the deep end.

CHAPTER 4

The Brute Facts of the Big Bang

At this moment it seems as though science will never be able to raise the curtain on the mystery of creation. For the scientist who has lived by his faith in the power of reason, the story ends like a bad dream. He has scaled the mountains of ignorance; he is about to conquer the highest peak; as he pulls himself over the final rock, he is greeted by a band of theologians who have been sitting there for centuries.

—Robert Jastrow[1]

Sometime in the early 2000s, I did have a road-to-Damascus moment, but it was empirical rather than spiritual. I decided that the entity that I might as well call God had deliberately designed the universe to permit the existence of life. It was a case of reading a relatively small number of pages of text and saying to myself, "I can't believe I'm thinking this, but it's the only plausible explanation."

This is a good point at which to make my amateur standing explicit. In the rest of the book, I take on a series of complex and often controversial topics. When I write books about topics on which I consider myself an expert—IQ, for example, or social policy—I am tacitly claiming that my presentation can be trusted, that I know what I'm talking about. When I discuss the physics of the Big Bang, the nature of consciousness, or the reliability of the Gospels, my opinions are worth no more than those of any other layperson who has approached unfamiliar and complex topics. I've done a lot of homework and I'm presenting my considered judgments, but there's no reason for you to assume I'm right. If you, like me, are taking religion seriously for the first time, you face the same problem: You are forced to decide what you make of a wide variety of issues that you do not have the option of mastering. I'm humbly offering a personal example of how the process works and where I came out, not a set of conclusions you should accept on my say-so.[2]

Genesis Redux

In 1920, scarcely more than a century ago, virtually all astronomers still believed that the universe was static, eternal, and consisted of just the Milky Way. Its eternal existence was "a brute fact," in the phrase that scientists sometimes use for things that theory does not explain. Albert Einstein believed so strongly in a static and eternal universe that in 1917 he introduced a fudge factor into his theory of general relativity to block the implication of his original equations, that the universe was either expanding or contracting.

The Brute Facts of the Big Bang

Now imagine someone saying in 1920, "Actually, the universe began at a specific moment about fourteen billion years ago when space itself and everything in it burst into existence from an infinitely small and dense point." It would have been seen as a clumsy attempt to impose a scientific gloss on "God said, 'Let there be light' and there was light."

That dismissive reaction occurred when in 1927 Belgian physicist and astronomer Georges Lemaître proposed the initial version of what later became known as the Big Bang Theory. Since Lemaître was also an ordained Roman Catholic priest, some of his scientific colleagues thought that he had let his faith contaminate his science.

But by 1927 events were already moving too rapidly for the scientific community to ignore Lemaître's hypothesis. Three years earlier, American astronomer Edwin Hubble had proved that the so-called nebulae were other galaxies, thereby transforming science's understanding of the size and structure of the universe. Two years after Lemaître's essay, Hubble used the redshift in the light of galaxies (a form of the Doppler effect) to support Lemaître's theoretical prediction that galaxies are moving away from Earth at speeds proportional to their distance. The equation that Lemaître and Hubble independently derived is now known popularly as Hubble's Law and more precisely as the Hubble-Lemaître Law.

The debate between the advocates of a creation event and of a steady-state universe continued from the 1930s into the 1950s with neither side having a decisive scientific advantage. The most famous advocate for the steady-state universe, Fred Hoyle, coined the label "Big Bang" for his adversaries' theory during a BBC radio broadcast in 1949, intending it to be derisory.

The balance of evidence shifted toward the Big Bang theory in the late 1950s and early 1960s as new discoveries indicated that the universe had been much hotter and denser in the distant past than it is now. In 1964 astrophysicists Arno Penzias and Robert Wilson discovered cosmic microwave background radiation permeating the universe—the afterglow of the cooling of the universe from its original hot plasma of sub-atomic particles. This was close to conclusive new evidence. By the 1990s, the core proposition of the Big Bang theory—that space, time, and all the matter in the universe originated in a singularity—was no longer in serious dispute.

I probably first became aware of the Big Bang theory in the 1960s or 1970s, but whatever I had read didn't intrude on my unreflective agnosticism until, when I was working on the book that would become *Human Accomplishment*, I was struck by the unlikelihood of twentieth-century science producing a creation story so close to the poetic description in Genesis. At that point, I sensed (though I could not articulate it to myself) that the reality of a beginning to the universe had momentous implications for the existence of the entity I might as well call God. I didn't read an articulation of that what I had sensed until I was working on this

The Brute Facts of the Big Bang

chapter and came across a passage in Stephen Meyer's *Return of the God Hypothesis*:

> If sometime in the finite past, either the curvature of space reached an infinite and/or the radius and spatial volume of the universe collapsed to zero units, then at that point there would be no space and no place for matter and energy to reside. Consequently, the possibility of a materialistic explanation would also evaporate, since at that point neither material particles nor energy fields would exist. Indeed, since matter and energy cannot exist until space (and probably time) begins to exist, a materialistic explanation involving either material particles or energy fields—before space and time existed—makes no sense.[3]

In other words, the Big Bang gave me good reason for thinking that the creation of the universe was a Mystery with a capital *M*. But that was still just preparing the ground for my road-to-Damascus revelation. The fantastic brute facts of the Big Bang forced me to rethink everything.

A Universe Created to Permit Life

Some Basics

The Big Bang was not an explosion that occurred in empty space. Space itself was created by the Big Bang. Current estimates of the chronology of the Big Bang involve unimaginably brief phases. The four fundamental forces—gravity, strong nuclear force, weak nuclear force, and electromagnetic force—are believed to have been unified as one until about 10^{-43} seconds after the Big Bang, when gravitational force emerged from the conglomeration. The strong nuclear force emerged at 10^{-36} seconds, and the weak nuclear force and electromagnetic force at 10^{-12} seconds. Or to put it another way, all these events are thought to have occurred in the first trillionth of a second.

The size of the universe increased unimaginably quickly. The initial inflation is thought to have occurred between 10^{-36} seconds and 10^{-32} seconds, when the universe went from unmeasurably small to a radius of several light-years.

The temperatures involved are also unimaginable. The temperature of the universe at 10^{-36} seconds is currently estimated to have been 10^{27} kelvin. Think of it this way: The surface temperature of our Sun is 5,500 degrees Celsius. With that in mind, try to try to imagine a trillion degrees. The temperature at 10^{-36} seconds after the Big Bang is estimated to have been a trillion times higher than that. As I said, unimaginable.

The Big Bang did not spew out the finished materials for two hundred billion galaxies. By about fifteen minutes after the Big Bang, hydrogen and helium-4 had formed, along with deuterium. The spherical volume of space was about three hundred light-years in radius. The density of matter was extremely thin—less than sea-level air density on today's Earth.

Now jump ahead some four hundred million years. The process that will eventually make life possible has started—the formation of stars. Gravity has caused matter to coalesce into clumps that will eventually become galaxies. Gravitational attraction also leads to clusters of hydrogen and helium molecules. The density of the clusters increases and their temperatures increase correspondingly. Eventually—we're talking about more millions of years—nuclear fission begins. After hundreds of millions or billions more years, stars start to burn out, generating heavier elements within their cores. When these dying stars explode into supernovae, the elements within the core are flung into space. After more hundreds of millions of years, the recoalescence of materials leads to the formation of second-generation and third-generation stars like our Sun—and, in some cases, planets orbiting those stars. By now the periodic table of naturally occurring elements has been filled, providing the raw material for you and me. As Francis Collins put it, "Nearly all of the atoms in your body were once cooked in the nuclear furnace of an ancient supernova—you are truly made of stardust."[4] It didn't have to work out that way. On the contrary, it is incredibly unlikely that it would work out that way. Therein lies the weirdness of the Big Bang.

> **SOURCES FOR THE ANTHROPIC PRINCIPLE**
>
> The formal name for the life-enabling nature of the Big Bang is the *anthropic principle*, first coined by cosmologist Brandon Carter at a conference in 1970 in a paper titled "Large Number Coincidences and the Anthropic Principle in Cosmology." Steve Weinberg's *The First Three Minutes: A Modern View of the Origin of the Universe* (1977) was an early (and excellent) introduction to the discoveries that verified the Big Bang theory. Martin Rees's *Just Six Numbers: The Deep Forces That Shape the Universe* (1999) is an accessible introduction to the mathematics of the Big Bang for lay readers.
>
> You have a choice of books that discuss evidence for the anthropic principle as it relates to religion. Two that I have read are Francis Collins's *The Language of God: A Scientist Presents Evidence for Belief* (2006) and Stephen Meyer's, *Return of the God Hypothesis: Three Scientific Discoveries That Reveal the Mind Behind the Universe* (2021). Others are Paul Davies's *The Mind of God: The Scientific Basis for a Rational World* (1992), Russell Stannard's *The God Experiment: Can Science Prove the Existence of God?* (2000), and Peter J. Bussey's *Signposts to God: How Modern Physics & Astronomy Point the Way to Belief* (2016).

Martin Rees's Just Six Numbers

I will use Martin Rees's book to explicate the weirdness. He wasn't the first to notice, and the physics surrounding the issue continued to evolve after he published in 1999, but *Just Six Numbers* is the book I read in the early 2000s that compelled me to accept that the universe was fine-tuned to permit life. Rees, then Britain's Astronomer Royal and former president of the Royal Society, had for me the added advantage of not being religious. His presentation had the credibility of a hostile witness giving favorable testimony.

Here are the major improbabilities that Rees discusses: Gravity is a feeble force. The electrical force that holds atoms together is vastly stronger than gravity—about 10^{36} times as strong. If gravity were even slightly stronger—for example, if the electrical forces that hold atoms together were only 10^{30} times stronger than gravity—then galaxies would form much more quickly and would be so densely packed that close encounters would be frequent, precluding stable planetary systems. In addition, a typical star would run out of fuel in about ten thousand years, exhausting itself long before life had a chance to evolve on an orbiting planet.

The stars are powered by the conversion of hydrogen (1 in the periodic table) into helium (2 in the periodic table). The fusion process converts a tiny portion of hydrogen's mass—to be specific, 0.007 of its mass—into energy. Life could not exist if that figure were much weaker or stronger. If the value were 0.006, a proton could not bond to a neutron, helium could not be formed, and the entire process that produces the periodic table of elements would be foreclosed. If the value were 0.008, two hydrogen protons would be able to bind directly, and no hydrogen would remain. No hydrogen would mean no fuel for stars.

The density of the universe divided by the critical density (which is an exact balance between the rate of expansion and gravitational force) is approximately one—in mathematics, *unity*. If the ratio were much greater than unity, the universe would have collapsed before enough time had elapsed for the process that led to filling out the periodic table. If the ratio, known to physicists as Ω (the Greek symbol for omega), were much less than unity, the universe would have expanded so rapidly that galaxies and stars wouldn't have had a chance to form. How small was the tolerance for error? Here's how Rees described it:

> It's like sitting at the bottom of a well and throwing a stone up so that it just comes to a halt exactly at the top—the required precision is astonishing: at one second after the Big Bang, Ω cannot have differed from unity by more than one part in a million billion (one in 10^{15}) in order that the universe should now, after ten billion years, be still expanding and with a value of Ω that has certainly not departed wildly from unity.[5]

In 1998, astrophysicists confirmed that the rate at which the universe is expanding is not decreasing, as it should do if all the energy for the expansion was produced by the Big Bang. Rather, the rate at which galaxies are receding from each other is increasing—some sort of mysterious "anti-gravity force" is at work. More than a quarter of a century later, astrophysicists are still trying to understand what produces this force, but it has been determined that the ratio of the energy density of this mysterious force and the critical density necessary for a flat universe has a value of 10^{-5}, far weaker than gravity. If the ratio were much larger than 10^{-5}, the anti-gravity force would have overwhelmed gravity during the higher-density stages of the early universe, precluding the formation of galaxies.

The Brute Facts of the Big Bang

In his final example, Rees goes back to the initial inflation in the first fraction of a second after the Big Bang. The natural expectation is that the universe would have expanded over billions of years into thinly spread dark matter and hydrogen and helium gas. No stars, therefore no complexity, and certainly no life. But as it turns out, there were extremely small perturbations at the outset of the expansion—"ripples" in the local density—that gave gravity a chance to form clusters of material. To see how small those perturbations were, imagine a planet the size of the Earth that is a perfectly smooth globe except for comparably small ripples. The tallest hill would be less than 200 feet high. The number expressing the ripples' amplitude works out to about 10^{-5}. It needs to be that magnitude, neither much smaller nor much larger. If it were much smaller than 10^{-5}, star formation would be slow and inefficient, and when the stars died their material would be blown out of the galaxy rather than being recycled into new stars that could form planetary systems. If the ratio were much larger than 10^{-5}, the universe would be a turbulent place with regions far bigger than galaxies that would not fragment into stars but collapse into vast black holes.

Apart from the numbers, Rees points out the unlikelihood that any matter at all existed after the Big Bang. Every kind of particle has an antiparticle. Particles and antiparticles annihilate when they encounter one another, and their energy is converted into radiation. The parsimonious expectation for the Big Bang is that it would have contained equal numbers of particles and antiparticles, all protons and antiprotons would have annihilated during the dense early stages, and the universe would have ended up full of radiation but no atoms, no stars, and no galaxies.

Instead, a small asymmetry favored particles over antiparticles, creating a slight excess of quarks over antiquarks. The asymmetry amounted to one extra quark for every 10^9 quark-antiquark pairs. As the universe cooled, antiquarks all annihilated with quarks, eventually giving quanta of radiation. But one out of every billion quarks survived because it couldn't find a partner to annihilate with. This is consistent with the empirical observation that radiation quanta in the universe outnumber protons by a billion to one. So all the atoms in the universe could result from a tiny bias at the ninth decimal place in favor of matter over antimatter. Theory provides no explanation why such a serendipitous asymmetry might exist.

I have given you a highly abbreviated overview of the fine-tuning that Rees described in *Just Six Numbers*, which in turn constitutes only a small portion of the total instances of fine-tuning described in the other books I list in the box.

Roger Penrose, Nobel laureate in physics and co-creator of the Penrose-Hawking singularity theorems, produced a summary expression of the unlikelihood of a universe that permits life by using the concept of entropy, which measures the amount of disorder in a material system.[6] The universe we inhabit constitutes a low-entropy, highly ordered arrangement of matter. Penrose set out to measure the probability that this would have occurred by chance. Mathematically, he was comparing the number of configurations associated with the maximum possible entropy state (a universe dominated by black holes) with the number associated with our low-entropy universe. The probability he came up with is 1 divided by a hyper-exponential number: 10 raised to the 10th power (10 billion) raised again to the 123rd power. The precise accuracy of that estimate is contentious, but there's plenty of room for error without changing the implications. Stephen Meyer put it this way: "If we tried to write out this number, there would be more zeros in the resulting number than there are elementary particles in the entire universe. I'm not aware of a word in English that does justice to the kind of precision we are discussing."[7]

THE MULTIVERSE ALTERNATIVE

There's a way around the extreme unlikelihood that our universe exists: The universe we inhabit could be just one of many. This alternative goes under the label of *multiverse theory*, with various approaches based on quantum mechanics, inflationary models that create "bubble" universes, and models based on string theory. The box gives you some books for educating yourself.

> **SOURCES FOR MULTIVERSE THEORY**
>
> Accessible sources advocating the multiverse theory are Brian Greene's *The Hidden Reality: Parallel Universes and the Deep Laws of the Cosmos* (2011), Michio Kaku's *Parallel Worlds: A Journey Through Creation, Higher Dimensions, and the Future of the Cosmos* (2002), and Sean Carroll's *Something Deeply Hidden: Quantum Worlds and the Emergence of Spacetime* (2020).
>
> A technical critical treatment of multiverse theory is Roger Penrose's *Fashion, Faith, and Fantasy in the New Physics of the Universe* (2016). More accessible critiques are Lee Smolin's *The Trouble with Physics: The Rise of String Theory, the Fall of a Science, and What Comes Next* (2006), Peter Woit's *Not Even Wrong: The Failure of String Theory and the Continuing Challenge to Unify the Laws of Physics* (2006), and Jim Baggott's *Farewell to Reality: How Modern Physics Has Betrayed the Search for Scientific Truth* (2013).

I can't help you. I am not competent even to describe the hypotheses, let alone make technical judgments about them. Rees, writing in 1999, was enthusiastic about pursuing these speculations because he judged that doing so falls within the proper domain of science—the speculations can be falsified given advances in our knowledge. The critics argue that the enterprise is a caricature of science. At the least, it can safely be said that investigations into multiverse theory have not yet yielded any empirical evidence for its validity.

My Options (and Yours)

After reading *Just Six Numbers*, I found myself limited to only three options:

The Brute Facts of the Big Bang

Option 1. One universe exists, it happens to permit life, and we lucked into an inexpressibly small chance.
Option 2. So many universes exist that it is not surprising that one of them, ours, permitted life to exist.
Option 3. One universe exists, and it was designed to permit life.

Which of the three is most plausible?

Regarding the first option, one uncurious reaction adopted by many scientists is to say there's no point in thinking about it. The brute fact is that we're here, and therefore our universe can support life. I prefer the metaphor proposed by Canadian philosopher John Leslie. Suppose that you faced a firing squad of a hundred expert riflemen and all of them missed. True, if they hadn't missed, you wouldn't be around to wonder why. But wondering why is nonetheless appropriate, because it is very odd indeed that all hundred missed.[8] For me, it's not even a close call between hypothesizing that all hundred accidentally missed versus hypothesizing that some unknown authority issued an order that everybody miss.

Regarding the second option, Roger Penrose's ratio tells me that millions of universes are required to have a reasonable chance that one of them permits the development of life. I cannot make myself take that option seriously. I am reminded of Samuel Johnson's response when Boswell asked him how he refuted George Berkeley's hypothesis that material objects do not exist independently of perception. "I refute it thus!" Johnson said, kicking a rock. When it comes to multiverse theory, "I refute it thus" by looking into a cloudless night sky.

That leaves the third alternative: I live in a universe that was intentionally designed to permit the development of life. That hypothesis has none of the drawbacks of the first two options. It requires me to assume an initial miracle that created something rather than nothing, but so do the other options! As far as I can tell, all the cosmologists' explanations about the origin of the universe are implicitly saying, "I'm going to ignore an initial miracle and explain everything else."

And so it is that I read *Just Six Numbers* and came away thinking that I had no choice. For me, the inescapable conclusion is that a God created a universe that would enable life to exist. Taking that step opened all sorts of other possibilities.

CHAPTER 5

Challenges to Materialism

One of the main reasons that "smart people don't believe that stuff anymore" is the widespread conviction among smart people that consciousness exists entirely and exclusively in the brain. I certainly shared that conviction from the time I went to college through my fifties. I was an orthodox materialist. It seemed so obvious: The brain is the only *possible* source of consciousness. Since that's true, life after death is impossible. Since all religions include belief in an afterlife (albeit sometimes vaguely, as in the case of Judaism), all religions are wrong.

That "it seemed so obvious" is an indication of how thoroughly college had socialized me to be a child of the Enlightenment. People throughout the world and throughout history until the Enlightenment had assumed the opposite: Humans possess souls that exist independently of the brain.

If you had known me really well, you could have chided me for being self-contradictory. I had been a materialist since college, and yet I had been intrigued by evidence of paranormal phenomena since early adolescence, and that interest had persisted. As an adult, I thought that evidence for some kinds of psychic phenomena, including telepathy and distance seeing, was strong. Shouldn't that have given me reason to doubt that science had produced all the answers about humans and souls?

Technically, materialism and acceptance of some types of paranormal phenomena can be reconciled. One can imagine explanations involving as-yet-undiscovered capabilities of the brain. But the tension between paranormal phenomena and materialism is undeniable. In recent decades, that tension has developed into full-blown conflict. Evidence that consciousness can exist independently of the brain needs to be taken more seriously than the scientific establishment has been willing to do. I'll start with the ways in which humans acquire and transmit information, then turn to evidence that consciousness can exist without the brain.

The Acquisition and Transmission of Information

William James was a founder of the philosophical position known as pragmatism, a notably hard-headed way to look at the world. He was also one of the leading American students of paranormal phenomena. He saw no contradiction between those positions. "Anyone with a healthy sense for evidence," he wrote, "a sense not methodically blunted by the sectarianism of 'Science,' ought now, it seems to me, to feel that exalted sensibilities and memories, veridical phantasms, haunted houses, trances with supernormal faculty, and even experimental thought-transference, are natural kinds of phenomena which ought, just like other natural events, to be followed up with scientific curiosity."[1]

James wrote that passage in 1903, but he had been interested in paranormal phenomena since the 1870s. He was joined in his curiosity by some of the most eminent British scientists of that age. In 1882, they founded the Society for Psychical Research.

They were immediately and ferociously attacked by the scientific establishments in both Britain and the United States. Paranormal phenomena implied a supernatural force. If the Enlightenment had accomplished anything, surely it had established the ascendancy of Science over God.

The Society for Psychical Research began its work in the heyday of mediums and seances, with their mysterious clicks, rappings, trumpet blasts, hazy apparitions in darkened rooms, and levitating tables. Many of the Society's successes involved the unmasking of the frauds. But key members of the organization soon realized that a much richer source of evidence was the case study—episodes of clairvoyance, apparitions, telepathy, or precognition as reported by ordinary people and accompanied by documentation and witnesses suitable for investigation by members of the society.

In the 1930s, Joseph Rhine, a professor at Duke University, began the quantitative study of extrasensory perception under experimental laboratory conditions. By the 1960s, quantitative research into telepathy, clairvoyance, precognition, and psychokinesis, now known collectively as *psi phenomena*, had gone beyond Rhine's famous flash cards. The methodological weaknesses and occasional fraud in the early research had been recognized and countered with better methods, and the subsequent sixty years has seen the accumulation of an extensive body of evidence.

I think that body of work has demonstrated the statistical reality of psi phenomena. The problem has been to get mainstream science to take the evidence seriously, because psi phenomena are so inherently unsuited to controlled laboratory conditions. The eminent physicist Freeman Dyson put it this way:

> If one believes, as many of my scientific colleagues believe, that the scope of science is unlimited, then science can ultimately explain everything in the universe, and ESP [extrasensory perception] must either be nonexistent or scientifically explainable. If one believes, as I do, that ESP exists but is scientifically untestable, one must believe that the scope of science is limited. I put forward, as a working hypothesis, that ESP is real but belongs to a mental universe that is too fluid and evanescent to fit within the rigid protocols of controlled scientific testing. I do not claim that this hypothesis is true. I claim only that it is consistent with the evidence and worthy of consideration.[2]

Challenges to Materialism

SOURCES FOR PARANORMAL PHENOMENA

You can get a sense of the scope of the Society for Psychical Research's early efforts by downloading the eBook version of *Phantasms of the Living* (1886), a two-volume work totaling more than 1,500 pages.[3] For William James's own views, see *William James on Psychical Research*, edited by Gardner Murphy and Robert Ballou (1960). For a fascinating account of the chief figures and their work during the first decades of research into paranormal phenomena, see Deborah Blum's *Ghost Hunters: William James and the Search for Scientific Proof of Life After Death* (2006).

For a sense of the modern body of quantitative psi research, two thick compendia are available: *Parapsychology: A Handbook for the 21st Century*, edited by Etzel Cardeña, John Palmer, and David Marcusson-Clavertz (2015) and *Evidence for Psi: Thirteen Empirical Research Reports*, edited by Damien Broderick and Ben Goertzel (2014). Both volumes contain experimental studies that are methodologically and statistically sophisticated.

The best one-volume summary of the evidence I have found is Dean Radin's *The Conscious Universe: The Scientific Truth of Psychic Phenomena* (1997). For accounts of many of the most convincing case studies, I recommend Elizabeth Mayer's *Extraordinary Knowing: Science, Skepticism, and the Inexplicable Powers of the Human Mind* (2007).

For critiques of the evidence for paranormal phenomena, see Benjamin Radford's *Scientific Paranormal Investigation: How to Solve Unexplained Mysteries* (2010) and his *Investigating Ghosts: The Scientific Search for Spirits* (2018), Richard Wiseman's *Paranormality. The Science of the Supernatural* (2015), and the bestseller by Carl Sagan, *The Demon-Haunted World: Science as a Candle in the Dark* (1996). Two magazines critical of paranormal phenomena are *Skeptical Inquirer*, published by the Committee for Skeptical Inquiry (formerly the Committee for the Scientific Investigation of Claims of the Paranormal), and *Skeptic*, the quarterly magazine of The Skeptics Society, founded by Michael Schermer and Pat Linse.

In my view, Dyson's phrase "scientifically untestable" is too sweeping. Controlled scientific testing has provided rigorous quantitative evidence for answering "yes" to the binary yes-no question, "Do any psi phenomena exist?" That finding has a huge implication all by itself: Human beings can acquire and transmit information without employing the five traditionally defined senses of sight, hearing, smell, taste, or touch.[4] However, I agree with Dyson that the most interesting evidence of paranormal phenomena, and the examples that are our basis for making serious progress, occur outside the laboratory, often under peculiar conditions involving extreme stress.

None of the sources listed in the box above directly addresses evidence that consciousness can exist independently of a functioning brain. For that, we must turn to two other types of paranormal phenomena: near-death experiences and terminal lucidity.

Near-Death Experiences

You have probably heard of the phenomenon known as the near-death experience (NDE). First brought to public attention by Raymond Moody in his 1975 book, *Life After Life*, NDEs have subsequently been the subject of books, articles in technical journals, and mass-circulation newspapers and magazines, plus a few movies. They have also been the subject of extensive scientific investigation, including the compilation of databases with thousands of cases.

Challenges to Materialism

An NDE occurs when death is impending or after the heart has stopped and a person is clinically dead. Some common characteristics are an awareness of being dead, an out-of-body experience, a life audit, a sense of peace and well-being, seeing and communicating with dead relatives and friends, and being in darkness (sometimes a tunnel) followed by immersion in a powerful light that is associated with love and unprecedented understanding. The obvious implication of NDEs is that some form of consciousness continues after biological death, contradicting the materialist position.

If you have read only popular accounts of NDEs, you may read into more rigorous discussions through the sources listed in the box below. Six aspects of the evidence impress me:

▸ In many well-documented cases, the person who had an NDE (called an "experiencer") acquired accurate factual knowledge during the NDE. In some cases, the experiencer remembered details of the procedures employed in resuscitating them that they should not have been able to describe, including idiosyncratic events—for example, where the nurse had put the experiencer's dentures. In other cases, experiencers have accurately reported events and conversations that occurred outside the experiencer's immediate area.[5]
▸ After they recover, experiencers continue to believe that what happened to them was profound. Anyone who has been drunk or high on drugs

can recall having pseudo-profound insights that they realized were silly once they sobered up. Not so with people who have experienced NDEs. Such people also often report long-lasting effects on their behavior and lifestyle, as well as beliefs.
- The materialist attempts to explain NDEs (e.g., from oxygen deprivation, ketamine in the brain, temporal lobe disruption) are relevant to some NDEs, but they cannot explain the most thoroughly documented cases of NDEs under cardiac arrest.
- The commonalities across NDE reports (e.g., the light at the end of the tunnel) cannot be explained as an artifact of the publicity given to NDEs. Those commonalities were reported by Ralph Moody in *Life After Life* before experiencers could have had access to others' accounts, and many more recent cases have been experienced by people who had not heard of NDEs.
- It is difficult to concoct an evolutionary explanation for NDEs. Psychological events that occur only while people are dying cannot contribute to reproductive fitness. Any evolutionary explanation must invoke some trait that contributes to reproductive fitness *and* presents as an NDE when the person is dying. I have not seen a plausible candidate.
- The state of knowledge about NDEs now includes prospective studies in which all occurrences of cardiac arrest in a set of hospitals were investigated in the first days following the

medical episode. Prospective studies with systematic protocols sidestep many of the problems associated with after-the-fact data collection.[6] More broadly, the science supporting NDEs has become increasingly sophisticated over the fifty years since Moody's *Life After Life* was published.

SOURCES FOR NEAR-DEATH EXPERIENCES

A recent source with the most thorough account of what has been learned about brain function near and at death is Sam Parnia's *Lucid Dying: The New Science Revolutionizing How We Understand Life and Death* (2024). Bruce Greyson's *After: A Doctor Explores What Near-Death Experiences Reveal About Life and Beyond* (2021) has extensive accounts of case studies. Pim van Lommel's *Consciousness Beyond Life: The Science of Near-Death Experiences* (2010) is the most comprehensive work, covering a variety of ancillary topics. If you're new to the subject, I recommend Judy Bachrach's *Glimpsing Heaven: The Stories and Science of Life After Death* (2014), an excellent journalistic account.

For articles critical of NDEs and other oddities of consciousness, see two collections of articles, *The Myth of an Afterlife: The Case Against Life After Death* (2015), edited by Michael Martin and Keith Augustine, and *Where God and Science Meet: How Brain and Evolutionary Studies Alter Our Understanding of Religion* (2006), edited by Patrick McNamara. For book-length critiques of NDEs, see Susan Blackmore's *Dying to Live: Near-Death Experiences* (1993) and *Near-Death Experiences: Understanding Visions of the Afterlife* (2016) by John Martin Fischer and Benjamin Mitchell-Yellin. Both Parnia's *Lucid Dying* and Greyson's *After* include references to more recent critical literature on NDEs.

I should add that while the NDE literature has religious implications, most people who have experienced an NDE do not dwell on these, nor do Christians consistently interpret the mystical figure in the dazzling light as Jesus. On the contrary, many devout Christians report that the specifics of Christian theology no longer seemed so important to them after their NDEs.

Terminal Lucidity

Terminal lucidity is the label for a sudden return to self-awareness, memory, and lucid functioning of a person who suffers from a severe neurological disorder that has deprived them of their mental capacities. The most common such disorder is advanced dementia, but terminal lucidity has also been reported among people suffering from meningitis, brain tumors, strokes, and chronic psychiatric disorders.

In the only systematic study with a large sample, described in Alexander Batthyány's *Threshold*, about 20 percent of the cases involved nonverbal communication (e.g., gestures, gaze) or verbal communication that was semi-coherent.[7] In the other 80 percent, people who had been unable to communicate anything were suddenly alert and "back" to their former personae. Terminal lucidity can last from minutes to a few hours. It is almost always followed by death within a day or so, with complete mental relapse in the interim.

Interpreting the data on terminal lucidity is simpler than interpreting the data on NDEs. Two major problems hinder interpretation of the portions of an NDE

Challenges to Materialism

SOURCES FOR TERMINAL LUCIDITY

A 2009 article by Michael Nahm and Bruce Greyson, "Terminal Lucidity in Patients with Chronic Schizophrenia and Dementia: A Survey of the Literature," reviewed more than eighty case histories, most of them clinical descriptions from the nineteenth century.[8] During the twentieth century, terminal lucidity continued to be observed in hospices, palliative care centers, and long-term care wards for persons with dementia, but it was usually treated as a curious episode that didn't warrant a write-up.

With the advent of social media in the twenty-first century, reports of terminal lucidity cases began to accumulate, and information about them began to be shared in a broader community. In 2008, a UN Panel titled "Beyond the Mind-Body Problem: New Paradigms in the Science of Consciousness" included a presentation on terminal lucidity.[9] In 2018, the National Institutes of Health sponsored a workshop on "Paradoxical Lucidity in Late-Stage Dementia," convening nine researchers from Europe and the United States ("paradoxical" was NIH's substitution for "terminal" but it did not catch on).[10]

Articles about terminal lucidity in technical journals have been appearing for the last fifteen years, but almost all in German. As I write, the only book-length English discussion of terminal lucidity is Alexander Batthyány's *Threshold: Terminal Lucidity and the Border of Life and Death* (2023).[11] It is a scientifically careful work containing detailed documentation.

---

that purportedly take place in an afterlife. First, we must take the experiencer's word for what happened. Researchers can verify an experiencer's report of an odd incident in the emergency room that took place while their heart was stopped, but not that a loving spiritual entity bathed in light stood at the end of a tunnel. Second, it is usually impossible to know the exact time when these experiences occurred, leaving open the possibility that some of them occurred during the recovery phase when brain function was returning.

Neither problem applies to terminal lucidity. Unless you think the material in *Threshold* is all an elaborate fabrication, four points seem secure:

- There's no uncertainty about when an episode of terminal lucidity occurs—it is observed "live."
- Observers are guaranteed. The only reason that anyone learns about an episode of terminal lucidity is that one or more people have been present when it happened. Nor are the observers limited to loved ones whose emotions or wishful thinking might affect the accuracy of their testimony. Physicians and other medical staff have observed terminal lucidity and recorded clinical details.
- Terminal lucidity has been observed for people whose brains are physiologically incapable of organized mental activity. Sometimes they have suffered extensive and documented brain damage impairing such critical regions of the brain as the hippocampus, frontal lobe, temporal lobe, or parietal lobe. Late-stage dementia typically involves atrophy of brain tissue, abnormal protein buildups, reduced neurotransmitter levels, and disrupted communication between neurons and across brain regions. These changes are irreversible. There is no possibility that terminal lucidity can occur because this damage has somehow been repaired.
- Many episodes of terminal lucidity occur for people who have not uttered a word or exhibited

any recognition of family members for many months or even years, and yet the evidence of lucidity is often incontrovertible, demonstrating the patient's possession of detailed memories, their ability to engage in logical conversation, and the return of personality characteristics.

It is difficult to reconcile these facts with the materialist position that consciousness cannot exist independently of brain function.

REINCARNATION

The evidence regarding reincarnation is another obvious source for exploring the existence of consciousness independently of the brain, but I do not include it here—though not for lack of evidence that children sometimes have quite specific memories, including names and events in a former life that have then been confirmed independently. The late Ian Stevenson was the acknowledged expert on the subject. For a one-volume overview of his decades of research, see Stevenson's *Children Who Remember Previous Lives: A Question of Reincarnation* (2000). You can find references to more recent studies on the website of the University of Virginia's Division of Perceptual Studies, which Stevenson founded.[10] I do not discuss that body of evidence because, as the title of Stevenson's book says, the evidence directly addresses the memories of children. That some children have such memories is as thoroughly established as the reality of near-death experiences, terminal lucidity, and other paranormal phenomena, but explaining those memories via reincarnation can be only an inference. In contrast to NDEs and terminal lucidity, there is no subset of cases that seem to exclude all explanations except consciousness independent of brain function.

Three Propositions

My account of the evidence for paranormal phenomena in this chapter should not be mistaken for wholesale acceptance. That evidence is notoriously riddled with wishful thinking, selective memories, false memories, shoddy methodology, and, of course, deliberate fraud. William James understood that when he joined the first attempts to examine paranormal phenomena scientifically. It was the reason for his famous pronouncement: "If you wish to upset the law that all crows are black, you must not seek to show that no crows are; it is enough if you prove one single crow to be white."[13] My conclusions are accordingly limited to three propositions.

The binary yes-no question, "Have psi phenomena been proven to exist?" has been answered "yes" with evidence that meets high standards of scientific proof. Denying that proposition requires throwing out extensive evidence from carefully controlled experiments not because it has been explained away, but because of the assumption that it *could be* explained away given enough investigation—a position that seems more akin to invincible faith than to carefully considered scientific judgment.

A subset of near-death experiences amounts to persuasive evidence that is incompatible with a strict materialist theory of consciousness. That subset consists of cases in which the experiencer was in full arrest (absence of cardiac output, absence of respiration, and absence of brain-stem reflexes), reported an out-of-body experience on site (e.g., in the operating room, emergency room, ambulance, or at the scene of the accident), and later accurately recounted an idiosyncratic event (e.g. an unusual resuscitation procedure, unusual apparatus, or specific actions or verbal exchanges of on-site personnel) that occurred while the experiencer was still in full arrest and that is independently verified.

A subset of terminal lucidity cases amounts to persuasive evidence that is incompatible with a strict materialist theory of consciousness. That subset consists of cases witnessed by more than one person, including at least one health-care worker not related to the subject who attested to the accuracy of the interchanges as reported by relatives, involving a subject who suffered from medically documented brain damage that excluded the possibility of organized thought and expression, and a subject who recognized relatives and friends and expressed memories of previous events.

Numerous cases meet these criteria for both NDEs and terminal lucidity. In my judgment they add up to proof that the materialist explanation of consciousness is incomplete. They are white crows.

In making that judgment, I lost what had been one of my sturdiest bulwarks against religious belief that involves me personally. If I am not just a brain in a body, what am I? I had to acknowledge the possibility that I have a soul.

PART II

TAKING CHRISTIANITY SERIOUSLY

CHAPTER 6

A Strange New Respect

Part I was about rethinking my agnosticism from the mid-1990s into the early 2000s. Part II is about gradually coming to accept many of the core tenets of Christianity over the subsequent twenty years. It probably pushes things too far for some of you. Evaluating the religious implications of contemporary science is one thing. Drawing conclusions about God's relationship to humans based on texts written two thousand years ago may strike you as foolish. But the material I describe in Part II poses intellectually intriguing puzzles, so you may want to give it a try anyway.

In the summer of 1997, I began work on the book that would be published six years later as *Human Accomplishment*.[1] I had two goals in mind. The first was to compile great accomplishments in the arts and sciences and the people who achieved them starting with the age of Homer and the Old Testament, circa 800 BCE, and continuing through the first half of the twentieth century. My second purpose was to explain *why* certain periods in certain places—Periclean Athens, Song China, late medieval Arabia, Renaissance Florence—had produced florescences of accomplishment.

One conspicuous florescence occurred in Western Europe from the fifteenth through the nineteenth centuries. Trying to explain the *why* of this creative explosion was one of the most intense periods of my work life. At the end of it, I had a strange new respect for Christianity.[2]

Christianity and Individualism

I began work on *Human Accomplishment* with some expectations about the importance of individualism. In an earlier book, I had taken Aristotle's discussion of human happiness in the *Nicomachean Ethics* as my framework.[3] One aspect was Aristotle's description of how humans enjoy themselves. In *A Theory of Justice*, John Rawls labeled it "the Aristotelian Principle" and summarized it this way:

> Other things equal, human beings enjoy the exercise of their realized capacities (their innate or trained abilities), and this enjoyment increases the more the capacity is realized, or the greater its complexity.[4]

This focus on realizing one's personal potential had no analogue in any East Asian, South Asian, Mesoamerican, or Arabic civilization before or since. To this extent, ancient Greece set the stage for modern Western individualism before Christianity came along.

But ancient Greece was not the kind of individualistic culture that the West would later become. Among other things, the ancient Greeks did not see Aristotelian happiness as something that everyone could enjoy. Only a small minority had the character and intelligence to strive for it.

Christianity was different. Like Judaism, it taught that individual human beings are invited into a personal relationship with God that is accessible to everyone regardless of their earthly station. But Judaism was also deeply grounded in community, with theology and practices emphasizing that the individual's spiritual life is lived within and for the community. As Christianity evolved away from communal life in its early centuries, the importance of the individual's salvation through a personal relationship with God grew. By the late Middle Ages, Christianity put the individual at center stage as no other philosophy or religion had ever done before. The potentially revolutionary message of Christian individualism was elaborated most influentially by Thomas Aquinas, who grafted a humanistic strain onto Christian theology, joining an inspirational message of God's love and his promise of immortality with an injunction to use all of one's individual capacities of intellect and will for the greater glory of God. It was a potent combination.

CHRISTIANITY AND SCIENCE

In *Human Accomplishment*, I argued that a major stream of accomplishment in the arts and sciences is fostered by a culture in which the most talented people believe that life has a purpose and that individuals can act efficaciously to fulfill that purpose. I gave credit to Thomist theology on both counts and extra credit to Protestantism for augmenting the individual's sense of efficacious autonomy. But when it came to science, I shared the widespread view that the Catholic Church had been institutionally hostile to science and that organized Christianity in general had been a drag on scientific research that conflicted with orthodoxy on such topics as the geological history of the Earth, evolution, and, more recently, homosexuality.

Then, in the summer of 2003, I bought a copy of Rodney Stark's newly published book *For the Glory of God*.[5] It was too late for me to make major revisions—*Human Accomplishment* was already in page proofs—so I had to make do with an endnote saying that *For the Glory of God* "makes a number of points that are coordinate with my arguments regarding purpose and autonomy."[6]

In reality, Stark had done much more, making a fully fleshed-out case for Christianity's unique role. Why did science develop in Europe and nowhere else? "My answer to this question is as brief as it is unoriginal," Stark wrote. "Christianity depicted God as a rational, responsive, dependable, and omnipotent being and the universe as his personal creation, thus having a rational, lawful, stable structure awaiting human comprehension."[7]

Stark included a passage that I wished had been in *Human Acccomplishment*. It was from a lecture at Harvard in 1925 by Alfred North Whitehead, coauthor (with Bertrand Russell) of *Principia Mathematica*. Referring to the scientist's "inexpungable belief" in the rationality and lawfulness of the universe, Whitehead wrote:

> When we compare this tone of thought in Europe with the attitude of other civilizations when left to themselves, there seems but one source of its origin. It must come from the medieval insistence on the rationality of God. . . . Every detail was supervised and ordered: the search into nature could only result in the vindication of the faith in rationality. *Remember that I am not talking of the explicit beliefs of a few individuals. What I mean is the impress on the European mind arising from the unquestioned faith of centuries. By this I mean an instinctive tone of thought and not a mere creed of words.*[8] [Emphasis added.]

I realized that *For the Glory of God* required me to rethink the relationship of organized Christianity to science. Greatly condensed, this is the story Stark recounts.

Even before Aquinas wrote in the thirteenth century, the Catholic church had produced the scholastics—Catholic intellectuals who were instrumental in starting the universities that evolved from Catholic cathedral schools established to train monks and priests. The universities of Bologna and Paris were opened around 1150, Oxford and Cambridge shortly after 1200, and then a flood of others during of rest of the thirteenth century. It is estimated that Europe's universities enrolled 750,000 students in their first 150 years.

These universities quickly became centers of higher learning in accordance with the teachings of Aquinas and other famous Catholic intellectuals of late medieval Europe—Albertus Magnus, William of Ockham, Jean Buridan, and Nicole Oresme among them. It wasn't only that Christian theology encouraged people to act purposefully and efficaciously as individuals. In a very real sense, the Scientific Revolution was sponsored by the Roman Catholic Church.

But what about the Church's opposition to Columbus because he thought the world was round? The burning of Giordano Bruno? The resistance to Copernicus's heliocentric solar system? The persecution of Galileo? Stark discusses each of these cases in detail over his seventy-seven-page chapter on science and religion, concluding with a damning accusation:

> The reason we didn't know the truth concerning these matters is that the claim of an inevitable and bitter warfare between religion and science has, for more than three centuries, been the primary polemical device used in the atheist attack on faith. From Thomas Hobbes through Carl Sagan and Richard Dawkins, false claims about religion and science have been used as weapons in the battle to "free" the human mind from the "fetters of faith."[9]

Read *For the Glory of God* and judge for yourself. It convinced me. Stark reads like an erudite scholar making a case that he realizes will attract criticism and that therefore requires abundant evidence. In my view, he met that challenge.

SOURCES FOR CHRISTIANITY'S ROLE IN WESTERN HISTORY

At least two books published since *For the Glory of God* expand on Stark's themes: Vishal Mangalwadi's *The Book that Made Your World: How the Bible Created the Soul of Western Civilization* (2012) and Tom Holland's *Dominion: How the Christian Revolution Remade the World* (2019).

CHRISTIANITY AND THE ARTS

One of the things that seemed most obvious to me as I compiled the inventories of great accomplishment in the arts was that the quality of the work fell off the edge of a cliff in the twentieth century. I can call upon scholarly support for that position from Jacques Barzun's magisterial *From Dawn to Decadence*.[10] I confess to personal animus as well. In the visual arts, I am baffled or bored by Dada, minimalism, conceptual art, pop art, performance art, and most (though not all) abstract expressionism. In music, I can barely make myself listen to the compositions of expressionism, serialism, or experimentalism. In literature, I find most postmodern novels to be lifeless.

So there you have it: I'm a philistine. But there is

another way of looking at it. For a variety of reasons, the classic ideals of the arts began to collapse toward the end of the nineteenth century.

In part, this phenomenon can be explained by a change in the way that artists saw their role. Catherine assigns much of the blame to the Romantic poets, who elevated the artist from craftsperson to a visionary figure—the "unacknowledged legislators of the world," in Shelley's words—and introduced the concept of the tortured artist.[11] I blame it on Beethoven, who was the exemplar of the rebellious, ill-tempered genius who breaks old rules and is contemptuous of his audience's preferences. He acted as if he were God's gift to humanity. As it happens, he was. The problem is that subsequent generations of artists who weren't gifts from God emulated him.

Artists' inflated self-importance combined with another change: secularism. Artists from the late nineteenth century onward were increasingly nonbelievers. In *Human Accomplishment*, I framed the effect of this change by using the three classic transcendental goods: Truth, Beauty, and the Good, each of which refers to perfect qualities that lie beyond direct experience.

From the late medieval period into the nineteenth century, conceptions of the True, Beautiful, and Good underwent changes. But if the question is *whether* artists saw themselves as aiming at the transcendental goods relevant to their respective fields, the answer is simpler: until the late 1800s, yes. Then, over a period starting in the late 1800s and extending through World War I, many of those who saw themselves engaged in high art consciously turned away from the idea that their function was to realize Beauty and then rejected the relevance of Truth and the Good as valid criteria for judging their work. Much "serious" fiction was written with a nihilistic, situational view of morality, and the pointlessness of life became a common theme. The visual artist's duty became to "challenge" the audience, from Marcel Duchamp's urinal to Andy Warhol's Campbell's soup cans to Andres Serrano's *Piss Christ*. In music, melody and harmony were redefined. Arnold Schoenberg, who announced the death of tonality and then did all he could to make his prediction come true, wrote that "those who compose because they want to please others, and have audiences in mind, are not real artists."[12] His contempt for the audience could not be plainer, nor could the godlike role that Schoenberg assigned to the artist.

All this led me to conclude that when religion no longer supplies a framework for thinking about the Good, True, and Beautiful, artists tend to make their work about their personal preferences, and those preferences tend to be banal, or wrongheaded, or both. That's one of the reasons that the high culture of the twentieth century seems so pallid in comparison with the high culture of the five preceding centuries.

By the time I had finished *Human Accomplishment*, I was convinced not only that religion had played a key role in igniting great accomplishment in the arts and sciences, but that Christianity had played a unique role. Secular people like me had to come to grips with that. "The easy answer is that the giants of the past were deluded," I wrote near the end of the book.

> They imagined that what they were doing had some transcendental significance, and, lo and behold, their foolishness inspired them to compose better music or paint better pictures. But this line of thought can become embarrassing when one confronts just what those self-deluded people accomplished. Is it not implausible that those individuals who accomplished things so beyond the rest of us just happened to be uniformly stupid about the great questions? Another possibility is that they understood things we don't. . . . Johann Sebastian Bach does not need to explain himself; he made a prima facie case that his way of looking at the universe needs to be taken seriously. It behooves us to do so.[13]

That last sentence encapsulates the difference between the person who began working on *Human Accomplishment* in 1997 and the one who finished the book in 2003.[14] I knew I should start taking Bach's way of looking at the universe seriously. But how?

CHAPTER 7

Enter C. S. Lewis

My daily log records that on April 27, 2005, I lunched with Pete Wehner at the White House Mess. Pete was then on President George W. Bush's staff. I had known him for years in his previous job as Bill Bennett's right-hand man when Bennett was Secretary of Education under Reagan and then Drug Czar in the first Bush White House. I had kept touch with him since then through an informal lunch group—Bill, Pete, Charles Krauthammer, and me—that had been convening at the Palm steakhouse three or four times a year since the early 1990s.

I knew that Pete was (and is) a devout Evangelical Protestant. Sometime during the lunch, I asked him how he had come to his faith. He replied that it was mostly because C. S. Lewis's *Mere Christianity* had convinced him of Christianity's truth. Within a few days I had bought a copy, read it, and been so impressed that I bought four more and gave them to my children—something I hadn't done with any other book.[1]

I am one of a multitude of readers who have reacted that way. *Mere Christianity* probably led more people to

Christianity than any other book of the twentieth century. How? For me, C. S. Lewis was the perfect example of a "smart person who still believed that stuff." His prose is informal, even chatty, but the man was an Oxford don, a classicist, and immensely erudite; his intellect shines through the deceptively simple points he makes. He also engages in a sort of dialogue with his reader, one of the things I enjoy most in a challenging work. You think you've found the weakness in his presentation, and in the next sentence he writes something to the effect that "Perhaps you're thinking. . . ." and responds to what you were thinking. Even when you decide that you don't agree, he has framed his argument in a way that periodically brings you back to reconsider it.

OTHER GATEWAY BOOKS TO CHRISTIANITY

Many, many introductions to Christianity have been written. Two authors I have read, Timothy Keller and Philip Yancey, might serve the function for you that C. S. Lewis did for me. I recommend Keller's *The Reason for God: Belief in an Age of Skepticism* (2008) and *Making Sense of God: Finding God in the Modern World* (2018) and Yancey's *What's So Amazing About Grace?* (1997) and *Soul Survivor: How Thirteen Unlikely Mentors Helped My Faith Survive the Church* (2003).

Two other books I have found to be especially illuminating, though perhaps they are not ideal entry-level books, are John Polkinghorne's *The Faith of a Physicist: Reflections of a Bottom-Up Thinker* (1994) and Cynthia Bourgeault's *The Wisdom Jesus: Transforming Heart and Mind—a New Perspective on Christ and His Message* (2008).

Mere Christianity engaged me in a mental dialogue with an excellent mind that accepted a Christian orthodoxy I had assumed could easily be dismissed. Two large issues stood out:

- What did I think of Lewis's argument about a Moral Law that originates with God?
- What did I think of Lewis's assertion that you do not have the option of saying that Jesus was a great moral teacher—that instead you are required to decide whether he was a lunatic, a liar, or the Son of God?

CHAPTER 8

The Moral Law

It's a big leap from accepting that the universe was designed to permit life to believing that God has anything to do with the lives of individual humans. C. S. Lewis made that leap in the first five chapters of *Mere Christianity* by arguing that our concepts of right and wrong come from God.

Lewis opens by talking about how people quarrel when they are upset at someone's behavior: "How'd you like it if anyone did the same to you?" "Leave him alone, he isn't doing you any harm." "Come on, you promised." In each case, Lewis points out, one person is appealing to a standard of decent behavior. "Quarrelling means trying to show that the other man is in the wrong," Lewis continues. "And there would be no sense in trying to do that unless you and he had some sort of agreement as to what Right and Wrong are; just as there would be no sense in saying that a footballer had committed a foul unless there was some agreement about the rules of football."[1]

Lewis postulates a Moral Law that human beings everywhere, across history and across cultures, are aware of even though they disobey it. But haven't different civilizations had radically different codes of right and wrong? Not really, says Lewis. The moral codes of the ancient Egyptians, Babylonians, Hindus, Chinese, Greeks, Romans, and the modern West have been remarkably similar, praising similar virtues and condemning similar sins.

> Think of a country where people were admired for running away in battle, or where a man felt proud of double-crossing all the people who had been kindest to him. You might just as well try to imagine a country where two and two made five. Men have differed as regards what people you ought to be unselfish to—whether it was only your own family, or your fellow countrymen, or everyone. But they have always agreed that you ought not to put yourself first. Selfishness has never been admired. Men have differed as to whether you should have one wife or four. But they have always agreed that you must not simply have any woman you liked.[2]

Lewis is asking nonbelievers to confront an elemental question: If moral codes are man-made, what is the authority for believing that a given act is wrong? For example, why are robbery, murder, and rape morally wrong? Whatever secular answers I come up with are vulnerable to a follow-up question: If the act has no bad consequences for *you*, and may even benefit you, why shouldn't you do it? Eventually I can be driven to the last-ditch secular defense: Society would disintegrate (and my life along with it) without rules against robbery and murder and rape—which is a roundabout way of saying that those acts are inexpedient.

I cannot get comfortable with that position. Perhaps you've run across the thought experiment that asks whether you would press a button that killed an anonymous person halfway around the world in return for a million dollars. Suppose we reword it. Would it be *wrong* to press the button? C. S. Lewis is saying that regardless of whether we would do it, we would know it was morally wrong. How about if the person were elderly, all his children were grown, and his wife had died? And think of all the socially useful things you could do with the million dollars. Those could be extenuating circumstances for someone who decides to push the button. They make the decision more understandable. But they don't alter the moral character of the act. It is wrong.

Why? I ultimately have no choice but to say that I think murdering someone is simply *wrong*. Lewis asks me to think about how that idea got so ineradicably into my head.

The obvious retort is that what Lewis calls the Moral Law reflects human instincts that have evolved over millions of years because they contribute to reproductive fitness. Lewis disagrees. "The Moral Law tells us the tune we have to play: our instincts are merely the keys." He makes three points:

First, if everything is a matter of evolved instincts, then the stronger of two competing instincts should govern our behavior. "But at those moments when we are most conscious of the Moral Law, it usually seems to be telling us to side with the weaker of the two impulses. You probably *want* to be safe much more than you want to help the man who is drowning: but the Moral Law tells you to help him all the same."

Second, the Moral Law often seems to be saying to us that we should make the weaker impulse stronger than it naturally is—that, for example, we should be less selfish and more generous. There is nothing logical in that internal sense unless we invoke a Moral Law that transcends instincts. "The thing that tells you which note on the piano needs to be played louder cannot itself be that note."

Third, if the Moral Law were merely a matter of instinct, we should be able to point to one of our instincts that is always in accord with the right thing to do. We cannot. Even when an instinct seems most uncomplicatedly good—for example, a mother's love for her child—that doesn't mean that all the behaviors that might be prompted by the instinct are good in all circumstances. "Think once again of a piano. It has not got two kinds of notes on it, the 'right' notes and the 'wrong' ones. Every single note is right at one time and wrong at another. The Moral Law is . . . something which makes a kind of tune (the tune we call goodness or right conduct) by directing the instincts."[3]

I found these arguments intriguing. But I was also convinced (and still am) that evolutionary psychology is one of social science's most important tools for understanding human behavior. Lewis wrote *Mere*

The Moral Law

Christianity in the 1940s, at a time when Darwinian evolution seemed to leave a big hole in explaining human behavior. Evolutionary selection pressures are blind to right and wrong. The only thing that matters is that one's genes get passed on to offspring. How could altruism have evolved in a world of the selfish gene? In the 1940s, evolutionary theory still had no well-developed answer.

I knew that it had subsequently been shown that evolution can produce altruistic behavior toward a biological relative when the fitness cost to the actor is smaller than the fitness benefit to the relative—*kin selection*.[4] It had also been shown that evolution can produce superficially altruistic behaviors among genetically unrelated individuals via *reciprocal altruism*, sometimes summarized as "I'll scratch your back if you scratch mine."[5] But neither of these discoveries explains what the ancient Greeks called *agape*: unconditional love, focused on giving rather than receiving, not based on merit or acquaintance with the recipient.

Theologically, *agape* has been used to denote God's love. Applied to humans, the most unambiguous manifestation of *agape* is kindness, concern, and help for complete strangers. Probably you have felt it yourself. As Francis Collins points out in *The Language of God*, "Surely most of us have at one time felt the inner calling to help a stranger in need, even with no likelihood of personal benefit. And if we have actually acted on that impulse, the consequence was often a warm sense of 'having done the right thing.'"[6]

Yes. In my case, I was driving down Massachusetts Avenue in Washington, DC, in the mid-1970s when a car a hundred yards in front of me collided with another car and burst into flames. My instantaneous twin reactions were "You have to stop and pull the driver out of the car" and, much louder in my head, "I DON'T WANT TO," fearing that the gas tank might explode at any moment. I did stop and, with another man who ran over, pulled the driver to safety. I have been relieved ever since that I did. Not stopping would have haunted me for the rest of my life even if someone else had rescued the driver. And in the moment, I did indeed have a warm sense of having done the right thing.[7]

The internal command to stop and help wasn't the result of any evolutionary mechanism that has yet been identified. All the known laws of evolutionary biology say that such behavior has a large negative effect on reproductive fitness. If not evolution, was the command the result of my parents' parenting? Lessons at Sunday school? Was it because I grew up in a sexist society in which males were indoctrinated to exhibit courage? I cannot prove that it was not caused by those things, but, fifty years later, I still remember how imperious that command was. It did not leave me an acceptable alternative. It felt a lot like a demand to live up to the Moral Law.

The tendency of human beings to behave altruistically is not limited to life-and-death situations. The odd thing about humans, once you stop to think about it, is how very common altruistic behaviors are. Generosity to strangers and engagement in charitable activities on behalf of strangers have been commented upon by visitors to this country since America was founded. The effects of communist regimes in stifling altruism are equally well documented. But whether the culture reinforces or suppresses it, genuine altruism that is not explained by evolutionary biology seems to be a fundamental component of human psychology.

So far, so good. C. S. Lewis, reinforced by Francis Collins, presented me with a powerful argument for the Moral Law. The more I have thought about it, the stronger their position appears. And yet I continued to struggle to accept the validity of the Moral Law wholeheartedly because the implications are so momentous. Lewis grouped the five chapters of *Mere Christianity* discussing the Moral Law under the title "Right and Wrong as a Clue to the Meaning of the Universe." In the fifth chapter, he finally reveals what he has in mind by that title. Here is the key passage, one that Francis Collins found so compelling that it marked a turning point in his journey to Christian faith:

> If there was a controlling power outside the universe, it could not show itself to us as one of the facts inside the universe—no more than the architect of a house could actually be a wall or staircase or fireplace in that house. The only way in which we could expect it to show itself would be inside ourselves as an influence or a command trying to get us to behave in a certain way. And that is just what we do find inside ourselves. Surely this ought to arouse our suspicions?[8]

Talk about a big leap. If I accept Lewis's and Collins's position unreservedly, I am driven to accept not only that God has a relationship with humans, but that God is ultimate Goodness. In the words of the first Epistle of John (4:7-8), "Beloved, let us love one another, because love is from God; everyone who loves is born of God and knows God. Whoever does not love does not know God, for God is love." And if I'm going to take that seriously, I must also contemplate the next sentence: "God's love was revealed among us in this way: God sent his only Son into the world so that we might live through him" (4:9).[9]

Which leads inexorably to my next task: to decide what I make of Jesus of Nazareth.

CHAPTER 9

Who Wrote the Gospels and When?

Catherine's return to Christianity began with an epiphany: Her love for her daughter was partly a conduit for a larger and transcendent love that pointed to God. Then and thereafter, the specifics of Christian theology have not been of much concern to her. Questions about the historicity of the New Testament are mildly interesting but beside the point.

For me, *Mere Christianity* posed a challenge. By 2005, I was already disposed to accept the existence of a God that was involved in some way with human life, and I could not dismiss Lewis's case for the Moral Law. But the rest of *Mere Christianity* was specifically about Christianity, and Lewis pulled no punches. The most in-your-face example is his famous "trilemma": Jesus was a liar, a lunatic, or the Son of God:

> I am trying here to prevent anyone saying the really foolish thing that people often say about Him: 'I'm ready to accept Jesus as a great moral teacher, but I don't accept His claim to be God.'

That is the one thing we must not say. A man who was merely a man and said the sort of things Jesus said would not be a great moral teacher. He would either be a lunatic—on a level with the man who says he is a poached egg—or else he would be the Devil of Hell. You must make your choice. Either this man was, and is, the Son of God: or else a madman or something worse. You can shut Him up for a fool, you can spit at Him and kill Him as a demon; or you can fall at His feet and call Him Lord and God. But let us not come with any patronizing nonsense about His being a great human teacher. He has not left that open to us. He did not intend to.[1]

There are many reasons to think that Lewis might be wrong about our having just three choices. For example, did Jesus really claim to be the Son of God or are those passages later inventions by the authors of the Gospels? Or perhaps there's a translation problem, or we just don't understand what he meant by language that seems to imply divinity. Those are reasonable objections. But Lewis's trilemma forces you defend your alternatives. It's not enough to assert that Jesus's claims to be the Son of God are later inventions; you had better read into the extensive literature about when the Gospels were written, by whom, and how trustworthy they are. It's not enough to say that Jesus didn't mean he was literally the Son of God; you had better read into still another literature dealing with that ambiguity.

And so I started reading, asking myself whether the New Testament and more specifically the Gospels—Matthew, Mark, Luke, and John—offer an interpretable account of the life and teachings of Jesus or amount to folklore. I cannot claim to have felt compelled to do so by religious fervor. I was just really curious.

THE REVISIONIST CHALLENGE

My understanding when I began was that the folklore view had won out. I was already familiar with something called the "Jesus Seminar." As I understood it, the members of the Jesus Seminar were rigorous New Testament scholars who were pooling their knowledge to identify the words in the Gospels that can confidently be attributed to Jesus, and their work had discredited the historical reliability of the Gospels. I was also aware of a popular analogy that was coming into use among these scholars, based on the children's game of telephone. Here is how it was explained by Bart Ehrman, the author of a widely used textbook on the early church:

> [N]early all of these storytellers had no independent knowledge of what really happened. It takes little imagination to realize what happened to the stories. You are probably familiar with the old birthday party game "telephone." A group of kids sits in a circle, the first tells a brief story to the one sitting next to her, who tells it to the next, and to the next, and so on, until it comes back full circle to the one who started it. Invariably, the story has changed so much in the process of retelling that everyone gets a good laugh. Imagine this same activity taking place, not in a solitary living room with ten kids on one afternoon, but over the expanse of the Roman Empire (some 2,500 miles across), with thousands of participants.[2]

My understanding was reasonably accurate, but I was soon to learn that the evolution in New Testament scholarship leading to the Jesus Seminar had a much longer history than I had realized.

In the second half of the eighteenth century, Enlightenment scholars began to approach the New Testament as they would any other ancient document, confident that Newton's clockwork universe left no room for the supernatural. Over the next two centuries, the revisionist search for the historical Jesus advanced progressively more ambitious claims that the Gospels were anonymous creations, contaminated by layers of alterations, and that the canonical four Gospels were arbitrary choices from a larger literature about Jesus that often contradicted them.

My term *revisionist* lumps together several different schools of thought. The early revisionists questioned whether the Gospels were authentically the work of their ascribed authors and whether the versions that survived were faithful to the originals. The last decades of the nineteenth century saw a series of German scholars who created "lives of Jesus" stripped of supernatural elements which, as Albert Schweitzer noted at the time, seemed to produce profiles of Jesus that were consistent with the secular ideologies of the scholars who created them.[3]

In the early twentieth century, a new generation of German scholars developed what came to be known as *form criticism*. *Form* refers to categories of biblical text such as miracles, parables, and sayings. The form critics set out to reconstruct the history of these forms before the Gospels were put in writing—the settings in which these forms were used, their literary structure and style, and how they were transmitted over time. This enabled them (they argued) to trace the development of Christian beliefs.

Around the middle of the twentieth century, form criticism was augmented by *redaction criticism*, which characterized the Gospel writers as anonymous editors who redacted and augmented the text depending on their own theological preferences. By the 1970s, the meaning of the plain text of the Gospels seemed to be of minor interest. Just as postmodernist literary criticism has often mirrored the late twentieth century's preoccupation with gender, race, and class, much of the deconstruction of the Gospels seems to have mirrored that era's skepticism of all religious tradition. By the end of the twentieth century, the received wisdom in the religion departments of many elite universities was that the New Testament is uninterpretable as history. This new received wisdom also seeped into the broader intellectual world, where people like me were predisposed to accept it.

REVISIONIST SOURCES

The revisionist sources I have used are John Dominic Crossan's *The Historical Jesus: The Life of a Mediterranean Jewish Peasant* (1991) and *The Birth of Christianity: Discovering What Happened in the Years Immediately After the Execution of Jesus* (1998); Burton L. Mack's *Who Wrote the New Testament? The Making of the Christian Myth* (1995); Charles Freeman's *A New History of Early Christianity* (2009); Robyn Faith Walsh's *The Origins of Early Christian Literature: Contextualizing the New Testament within Greco-Roman Literary Culture* (2021); Markus Vinzent's *Resetting the Origins of Christianity: A New Theory of Sources and Beginnings* (2023); and four books by Bart D. Ehrman: *The New Testament: A Historical Introduction to the Early Christian Writings* (1997), *Forged: Writing in the Name of God—Why the Bible's Authors Are Not Who We Think They Are* (2011), *How Jesus Became God: The Exaltation of a Jewish Preacher from Galilee* (2014), and *Jesus Before the Gospels: How the Earliest Christians Remembered, Changed, and Invented Their Stories of the Savior* (2016).

The Defense of Patristic Tradition

Then in 2007 I came across *Jesus and the Eyewitnesses: The Gospels as Eyewitness Testimony*, published a year earlier. The title grabbed my attention and I dived in. The book was more than five hundred pages long, dense with empirical material, and written by a respected British New Testament scholar, Richard Bauckham. Bauckham argued that the Gospels were consciously written to emphasize the body of eyewitness data contained in them and, more broadly, that the traditional understanding of the Gospels is much closer to the truth than the revisionists' accounts.

At the time, I thought Bauckham was a lonely voice. Being in the minority doesn't mean you're wrong (I know something about that), but usually there's a reason why the majority opinion has its majority. And yet as I read more about the revisionists, I was increasingly on Bauckham's side.

My reasons surely involve confirmation bias (the tendency to believe new information that supports what one already wants to believe), because even before finding *Jesus and the Eyewitness* I was becoming suspicious of the revisionists. What I had read of the revisionists' position was undeniably erudite, but often erudite in the same way that Jacques Derrida and Michel Foucault were erudite—meaning (in my view) that they devised convoluted explanations for phenomena that have simpler and more plausible explanations.

After reading deeper into the revisionists' work, I was also put off by their confidence that they could make judgments about the accuracy and trustworthiness of the patristic writers—the Church Fathers who wrote in the second and third centuries. Any modern New Testament scholar, traditional or revisionist, has access to only a fraction of the material available to the early Fathers. I didn't object to a critical reading of the materials that survive, but I felt that too often the revisionists blew off important information on tenuous grounds or too easily assumed that the patristic writers were ready to twist the facts to fit their agendas.

That, I thought, was exactly what some of the revisionists seemed to be doing.

I also discovered that I had overestimated Bauckham's loneliness. His *Eyewitnesses* was a significant contribution, but to an existing literature defending the traditional position against the revisionists' arguments. Since 2007, I have read more of both the traditional and revisionist literature than I had then, but my early opinions have been reinforced (confirmation bias again, you may reasonably suspect). The more I read, the more impressed I was by the empirical evidence and the logic deployed in defense of tradition, and the more exasperated I became with the revisionists.

I tell you all this so that you can read the rest of this chapter and the next with my priors on the table. I also ask you to recall what I said about my amateur standing: I'm giving my considered opinion, but you shouldn't assume I'm right. You need to do your own homework (this being part of it) and decide what you think.

The rest of this chapter takes up issues that are central to the revisionist position: the authorship of the Gospels and when they were written. Chapter 10 takes up the historicity of the Gospels. I devote so much space to the Gospels for two reasons. First, I want to give you a sense of the breadth and depth of the empirical arguments that have been brought to bear on the debate. You should not trust my conclusions, but you should realize how much material led me to them. Second, I found that satisfying myself that the Gospels are

Who Wrote the Gospels and When?

trustworthy has been invaluable. I am no longer prevented from absorbing what's in the text by worrying that it's all made up. Here I present material that might draw you into doing your own investigation, in hopes that you will reap the same reward.

SOURCES FOR THE HISTORICITY OF THE GOSPELS

The sources I am referencing on behalf of the traditional interpretation are, in order of publication, Larry Hurtado's *One God, One Lord: Early Christian Devotion and Ancient Jewish Monotheism* (1988); Paul Barnett's *The Birth of Christianity: The First Twenty Years* (2005), Richard Bauckham's *Jesus and the Eyewitnesses: The Gospels as Eyewitness Testimony* (2006), the second, revised edition of Craig Blomberg's *The Historical Reliability of the Gospels* (2007, first published in 1987), Michael Bird's *The Gospel of the Lord: How the Early Church Wrote the Story of Jesus* (2014), Brant Pitre's *The Case for Jesus: The Biblical and Historical Evidence for Christ* (2016), Lydia McGrew's *Hidden in Plain View: Undesigned Coincidences in the Gospels and Acts* (2017), and Peter J. Williams's *Can We Trust the Gospels?* (2018).

I do not discuss allegations, given wide publicity by Dan Brown's 2003 novel *The Da Vinci Code*, that the four canonical Gospels were chosen arbitrarily and other writings with equally good credentials were arbitrarily labeled heretical. If the issue interests you, several of the sources above discuss it, especially Bird's *The Gospel of the Lord*. For a book-length treatment, see Charles E. Hill's *Who Chose the Gospels? Probing the Great Gospel Conspiracy* (2010).

A note on usage. I usually refer to the Gospels as Matthew, Mark, Luke, and John, omitting "the Gospel According to . . ." for brevity. Conforming to usage in New Testament scholarship, I do not italicize these titles.

Who Wrote the Gospels?

The earliest known compilation of the four books now known as the Gospels dates to around 173 CE—almost a century and a half after the crucifixion of Jesus—but no copy of it survives.[4] The first surviving compilation (plus Acts), which is held at the Chester Beatty Library in Dublin, Ireland, is known as Papyrus 45. It dates to the first half of the third century.

What happened between the crucifixion and our earliest versions of the Gospels? The revisionists postulate anonymous manuscripts, augmented and redacted repeatedly and then arbitrarily assigned authorship that conferred authority on them. The alternative viewpoint is that the authorship of the four Gospels was well known throughout the early church and that these four accounts of the life of Jesus were widely accepted as authoritative because of the obvious credentials of the authors.

The Earliest Sources About Authorship of the Gospels
The earliest surviving testimony about the authorship of the Gospels comes from four men:

Papias (c. 60–130), Justin Martyr (c. 100–165), Irenaeus (c. 115–180), and Clement of Alexandria (c. 150–215). Much of the surviving material is found in Eusebius's *Ecclesiastical History*, which was completed in the early 320s and included verbatim excerpts of their work.

Papias, who eventually became Bishop of Hierapolis, is the only one of the four who had direct contact with the generation of the apostles and disciples (I use *apostles* exclusively to refer to the famous Twelve and *disciples* as a more general label that includes the many others who traveled with or were otherwise followers of Jesus during his ministry). Papias sought out information from those he called "the elders"—men who either were disciples or had been followers of the disciples in the first decades after the crucifixion. Many of them were still alive when he began assembling information in the 80s and 90s. Papias had an explicit preference for eyewitness accounts:

> I shall not hesitate also to put into properly ordered form for you everything I learned carefully in the past from the elders and noted down well, for the truth of which I can vouch. . . . And if by chance anyone who had been in attendance on the elders should come my way, I inquired about the words of the elders—that is, what Andrew or Peter said, or Philip, or Thomas or James, or John or Matthew or any other of the Lord's disciples, were saying. For I did not think that information from books would profit me as much as information from a living and surviving voice.[5]

Justin Martyr (martyred during the reign of Marcus Aurelius) was born in Neapolis in Samaria to pagan parents. He defined himself as a Gentile. After studying Stoicism and Pythagorean philosophy, he became a Christian, setting up his own school in Rome about 138.

Irenaeus, a Greek who eventually became Bishop of Lyons, had been a student of Polycarp, an early

Christian bishop. Irenaeus recalled how Polycarp "would tell of his conversations with John and with the others who had seen the Lord, how he would relate their words from memory; and what the things were which he had heard from them concerning the Lord, his mighty works and his teaching."[6] But it's not just his personal contact with Polycarp that makes Irenaeus an important source. He was a scholar who had access to early church writings that have since been lost.

Clement of Alexandria was an established scholar of Greek philosophy and literature before converting to Christianity. He then became a scholar of the early church. He was born too late to have had any personal contact with the disciples or elders, but he also had access to writings of the early Church Fathers that are lost to us, including the complete writings of Papias and Irenaeus.

Eusebius, who came a century later, was by reputation the greatest Christian scholar of his age. He was the literary heir of Origen and custodian of the library at Caesarea, the repository for the most complete documentation of the early church.

The Patristic Writers in Their Own Words
What are the actual words that these men used about authorship of the Gospels? I begin with an evocative passage by Clement of Alexandria about the apostles and why so few wrote down their memories.

> Those great and truly divine men, I mean the apostles of Christ, were purified in their life, and

were adorned with every virtue of the soul, but were uncultivated in speech. They were confident indeed in their trust in the divine and wonder-working power which was granted unto them by the Savior, but they did not know how, nor did they attempt to proclaim the doctrines of their teacher in studied and artistic language, but employing only the demonstration of the divine Spirit, which worked with them, and the wonder-working power of Christ, which was displayed through them, they published the knowledge of the kingdom of heaven throughout the whole world, paying little attention to the composition of written works. . . . Of all the disciples of the Lord, only Matthew and John have left us written memorials, and they, tradition says, were led to write only under the pressure of necessity.[7]

Now to the specifics of authorship:

The Gospel According to Matthew. The early Church Fathers, including several sources in addition to Papias, Irenaeus, and Clement, say that the apostle Matthew—a tax collector and probably the only literate apostle—wrote his account while he was still preaching in and around Jerusalem.[8] Papias writes that "Matthew composed the *logia* [sayings, oracles, and other scriptural text] in the Hebrew language, but each person interpreted them as best he could."[9] Irenaeus writes that "Matthew published a written Gospel among the Hebrews in their own language while Peter and Paul were preaching at Rome and founding the church."[10] Clement, who had noted that Matthew and John wrote their accounts out of necessity, went on to explain: "For

Matthew, who had at first preached to the Hebrews, when he was about to go to other peoples, committed his Gospel to writing in his native tongue, and thus compensated those whom he was obliged to leave for the loss of his presence."[11]

The Gospel According to Mark. The early Fathers agree that Mark recorded Peter's testimony. Papias quoted John the Elder as saying,

> Mark, having become Peter's interpreter, wrote down accurately everything he remembered, though not in order, of the things either said or done by Christ. For he neither heard the Lord nor followed him, but afterward, as I said, followed Peter, who adapted his teachings as needed but had no intention of giving an ordered account of the Lord's sayings. Consequently, Mark did nothing wrong in writing down some things as he remembered them, for he made it his one concern not to omit anything that he heard or make any false statement in them.[12]

In his surviving writings, Justin Martyr quotes from all four Gospels without naming the authors, referring to his sources as "the memoirs of the apostles and those who followed them," but he does refer explicitly to the "memoirs of Peter" in his quotation of Mark 3:17.[13]

Irenaeus wrote, "After their [Peter and Paul's] departure, Mark, the disciple and interpreter of Peter, did also hand down to us in writing the things what had been preached by Peter."[14]

Clement wrote that those who heard Peter preach

were not satisfied with a single hearing or with the unwritten teaching of the divine proclamation, but with every kind of exhortation besought Mark, whose Gospel is extant, seeing that he was Peter's follower, to leave them a written statement of the teaching given them verbally, nor did they cease until they had persuaded him, and so became the cause of the Scripture called the Gospel according to Mark. And they say that the apostle, knowing by revelation of the Spirit to him what had been done, was pleased at their zeal, and ratified the scripture for study in the churches.[15]

The Gospel According to Luke. All the early texts identify Luke as the Gentile disciple whom Paul refers to as "the beloved physician" in Colossians (4:14) and agree that he was Paul's frequent traveling companion. It is also agreed that Luke was the author of Acts of the Apostles and that Acts was written after Luke's Gospel.

The Gospel According to John. Unlike Matthew, who never says that he was an eyewitness, John appears to be explicit. In the concluding verses, he writes, "This is the disciple who is testifying to these things and has written them." John also refers several times to "the disciple Jesus loved," and traditional scholarship (though not revisionist scholarship) assumes the author was referring to himself.

Irenaeus writes that after the publication of the other Gospels, "then John, the disciple of the Lord, the one who leaned back on the Lord's breast, himself published a Gospel while he resided in Ephesus." He is alluding to John's narrative of the Last Supper.[16]

Clement writes that John's Gospel came about at the behest of friends who urged John to add material that the other three Gospels had omitted. "The three Gospels which had been written down before were distributed to all including himself [John]; it is said he welcomed them and testified to their truth but said that there was only lacking to the narrative the account of what was done by Christ at first and at the beginning of the preaching. . . . They say accordingly that John was asked to relate in his own Gospel the period passed over in silence by the former evangelists."[17]

The early and explicit statements of authorship are consistent. Whether all these early statements about authorship can be taken seriously is intensely controversial. The next chapter will take up my reasons for thinking they can.

When Were the Gospels Written?

The first modern attempt to date the books of the New Testament was Ferdinand Christian Baur's *Das Christenthum und die christliche Kirche der drei ersten Jahrhunderte* (1853), which argued that most of the New Testament dates from well into the second century. It was followed by J. B. Lightfoot's *Biblical Essays* (1893), which moved the dates up to the first century or the early years of the second century. Adolf von Harnack's *Die Chronologie der Litteratur bis Irenaeus*

Who Wrote the Gospels and When?

(1897)—Volume 2, Part 1, of his massive *Geschichte der altchristlichen Literatur bis Eusebius*—estimated dates that have been used by mainstream scholars since then. Harnack's dates put Mark first, at c. 70, Matthew at c. 80, Luke at c. 85, and John at c. 90.[18] These are roughly the dates that I took for granted in my unreflective years.

I recall being of two minds about the gap between Jesus's life and the first written accounts of it. The gap was long enough that memories could have blurred or been embellished, but it wasn't clear to me that the gap was long enough to give the telephone-game phenomenon time to introduce layers of major errors. The crucifixion is dated to 30 or 33 (both years are consistent with collateral evidence about the length of Jesus's ministry and the timing of Passover). Using the mainstream dates and ordering of the Gospels, the gap between Jesus's death and the appearance of Mark was a maximum of forty years. As of 2025, forty years ago is the beginning of Ronald Reagan's second presidential term. That may sound like ancient history if you're under sixty, but it is very much within living memory for those of us who are older.

It is also reasonable to expect that many participants in Jesus's ministry were still alive in the last decades of the first century. Overall life expectancy was much lower then, but that was driven by high infant and childhood mortality. The famous biblical reference to a lifespan of three score and ten was not unrealistic for people who had survived into adulthood. According to the conventional dating of the Gospels, many disciples who were eyewitnesses to Jesus's ministry and death could still have been alive when Matthew, Mark, and Luke were being written, which means they could have been used as sources. Just as important, many would have been alive to dispute false stories when the Gospels were first circulated—something that the authors of the Gospels had to be aware of.

Without giving it more thought than that, I wasn't convinced that the time gap in the conventional dating was automatically a big deal, but I did think that shorter would be better. What I didn't realize then is that the mainstream dating is on shaky ground. Three New Testament scholars have made a strong case that Matthew, Mark, and Luke were written no later than the 50s, with John following during the 60s, and that Matthew and Mark may have been written as early as the 40s.

The Basis for the Conventional Dating

"We may start with the fact, which I confess I did not appreciate before beginning the investigation, how *little* evidence there is for the dating of *any* of the New Testament writings." So writes John Robinson in the concluding chapter of *Redating the New Testament*.[19] Later, he quotes theologian Austin Farrer, who compares the conventional dates of the Gospels to "a line of tipsy revellers walking home arm-in-arm; each is kept in position by the others and none is firmly grounded."[20]

It's an apt way of describing the logic for the conventional dating. In 66 CE, Roman Emperor Nero dispatched an army to put down unrest in Judea. Four years of conflict led to a prolonged siege of Jerusalem and finally to the destruction of the Second Temple in 70. Matthew, Mark, and Luke all contain Jesus's prediction that the Second Temple will be destroyed. The conventional dating depends on the assumption that these were not predictions, but passages inserted after 70 to make it look as if Jesus foresaw the destruction. This supposition is also applied to other references in the Gospels that could possibly (though not definitely) reflect knowledge of events after 70. Together, they make up the slender foundation for dating the Gospels after 70.

The Case for Earlier Dates

The argument that the material about the destruction of the Temple was inserted after 70 has some problems. The first, of course, is that it implicitly rejects any validity for biblical prophecy. That's a reasonable Enlightenment position but doesn't apply if prophecy is a real phenomenon and especially if Jesus was the Son of God.

SOURCES FOR REDATING THE GOSPELS

The first book-length argument for an earlier dating of the New Testament was John Robinson's *Redating the New Testament* (1976). It was followed by John Wenham's *Redating Matthew, Mark and Luke: A Fresh Assault on the Synoptic Problem* (1992). The most recent work is Jonathan Bernier's *Rethinking the Dates of the New Testament: The Evidence for Early Composition* (2022).

The argument has problems even if prophecy is not a real phenomenon. The Jerusalem Temple had been destroyed previously, by Nebuchadnezzar in 586 BCE, and the language used in Jesus's prophecy includes details that apply to the Old Testament accounts of that invasion but not to General Titus's capture of Jerusalem in 70 CE. Jesus often drew material from the Old Testament and may have been doing so here. Also, certain parts of the text about the destruction of the Second Temple make sense only if they were set out *before* the event. For example, Mark and Matthew have Jesus saying that people should pray that the destruction "may not be in winter." Why would they write that if they already knew that the Temple was destroyed in summer? Another consideration: If the Gospels were written after the destruction of the Second Temple, it is odd that none of the Gospels mentions that Jesus's prophecy was fulfilled (as Luke does in reporting the fulfillment of Agabus's prediction of a famine in Acts 11:27–28).

I have used a few paragraphs to describe an issue that has been argued about in long book chapters by learned scholars. Here, similarly abbreviated, are the main arguments for early composition of the Gospels:

No Mention Anywhere in the New Testament That the Second Temple Was Destroyed. John Robinson turns the argument from the destruction of the Second Temple on its head. For Robinson, the significant clue, akin to Sherlock Holmes's dog that didn't bark in the nighttime, is that none of the books of the New Testament mentions that the Second Temple was destroyed. It was an epochal event in the history of Judaism, marking the transition of Judaism from a religion of the Temple to a religion of the Book. As of 70 CE, the nascent Christian sect still had a strong Jewish element. If the New Testament was indeed written after the destruction of the Second Temple, Robinson argues, it is inconceivable that none of its twenty-seven books would mention it, especially since it could be used as vindication of Jesus's prediction.

The Jewishness of the Gospels. Whoever the authors of the Gospels were, they or their sources were intimately familiar with the Old Testament and Jewish traditions.

Matthew begins with a genealogy of a type found throughout the Old Testament and contains more than three dozen quotations from the Old Testament. Mark begins with a quotation from the Old Testament, contains extended stories about Jewish theological debates, and is sprinkled with Jewish apocalyptic language. Jesus's last words from the cross, as reported in Matthew and Mark, "My God, my God, why have you forsaken me?" are a quotation from the opening of Psalm 22. John's opening echoes the beginning of the Old Testament. He subsequently exhibits his familiarity with the Jewish use of stone vessels for purification.

The Gospels display local knowledge of the Passover customs during Jesus's era, including the celebration of Passover within the walls of Jerusalem and the singing of the Hallel at the Passover feast. Matthew and Mark report that the high priest tore his clothes after judging Jesus to be guilty of blasphemy, a custom that rabbinic writings associate with a response to blasphemy.[21]

The Judaic roots of Christianity faded quickly as Christianity spread out of Palestine into Asia Minor, Greece, and Rome. The Jewishness of the Gospels suggests either that they were written by Jews or that the sources for the material were Jews and any augmentation and redaction managed to transmit that Jewishness. "We might put it like this," Peter Williams concludes. "The Four Gospels are so influenced by Judaism in their outlook, subject matter, and detail that it would be reasonable to date them considerably before the Jewish War. . . . Even if we say that the Gospels are late first century, the material in them is not."[22]

Internal Evidence for Pre-70 Origins of the Gospels. Biblical scholarship can call on two types of evidence: *external*, based on historical documentation from sources independent of the biblical text, and *internal*, based on analysis of the text itself. In his book on the evidence for early composition, Jonathan Bernier conducts an internal analysis, asking two questions of each of the books of the New Testament: Does the book contain material that is most fully intelligible only if written after 70? Does it contain material that is most fully intelligible only if written prior to 70? "Yes" to the first question points to post-70 composition regardless of the answer to the second question. Two "No" answers means that no conclusion can be drawn. "No" to post-70 material and "Yes" to pre-70 material mean that the book most likely predates 70. "It is the contention of this study," Bernier writes, "that while there are New Testament books that contain material most fully intelligible only if written prior to 70, there are no New Testament books that contain material most fully intelligible only if written after 70.[23] His analysis leads him to date Mark to 42–45 CE, Matthew to 45–59, Luke to 59, and John to 60–70.[24]

The Curious Ending of Acts. The most intriguing argument for early composition of the Gospels hangs upon the order in which the Gospels were written and the curious ending of Acts.

In the final chapter, Luke describes the end of Paul's voyage to Rome and Paul's meeting with Jewish elders in Rome three days after his arrival. Then comes the concluding sentence: "He lived there two whole years at his own expense and welcomed all who came to him, proclaiming the kingdom of God and teaching about the Lord Jesus Christ with all boldness and without hindrance." The end.

Paul's arrival in Rome is consensually dated to 60. He was martyred under Nero, probably in 64 during Nero's persecution of Christians after the Great Fire and no later than Nero's death in 68. Peter's death by crucifixion in Rome is dated to the same period. If Acts was written sometime between 70 and 85, when it is conventionally dated, why didn't the author mention Paul's martyrdom? Peter's too, for that matter?

If we accept the conventional dating, it's an extremely odd omission. The author could have concluded Acts with an account of the deaths of Paul and Peter, a landmark transition in the history of the new church. Such an account would be consistent with Luke's propensity to draw parallels between the life of Jesus and the lives of Paul and Peter. The author would also have been able to end Acts with an account of the survival and spread of the church after their deaths—a triumphant ending.

Instead, Acts ends abruptly in 62 with Paul living quietly in Rome and no mention of Peter. There is a natural explanation, one that applies to the biographies of people who are still living when the biography has been completed: There's nothing more to say. Luke, the author of Acts, finished writing Acts before Paul and Peter died—probably before 64—and he must have begun it at least a year or two earlier.

If you accept that logic, the dates for the writing of Mark, Matthew, and Luke are pushed much closer to Jesus's death. A consensus of scholars accepts that Luke is the author of both the Gospel According to Luke and Acts, and that he wrote the Gospel before he wrote Acts. And since it is also generally agreed that Luke had access to Matthew (or at least to the hypothesized source material for Matthew known as "Q") and to Mark when he wrote his Gospel, the necessary implication is that the Mark and Matthew Gospels (or at least Mark and "Q," including much of the Gospel of Matthew) were written no later than the 50s, only twenty to thirty years after Jesus's death, perhaps even in the 40s.

One other tidbit worth mentioning: If people kept augmenting and altering the books of the New Testament as the revisionists insist, why wouldn't someone have added a few lines to the ending of Acts mentioning the deaths of Paul and Peter? It's almost as if Luke finished his text before Paul and Peter were martyred and nobody subsequently messed with it.

The abrupt ending of Acts preyed so heavily on Adolph von Harnack, whose dates for the Gospels established the mainstream received wisdom, that he eventually changed his mind. Here is his description of what had happened to his thinking by 1911:

> Throughout eight whole chapters St. Luke keeps his readers intensely interested in the progress of the trial of St. Paul, simply that in the end he might disappoint them—they learn nothing of the final result of the trial! . . . One may object that the end of the Apostle was universally known, or one may also say that when the author had brought St. Paul to Rome he had attained the goal that he sets before himself in his book. For many years I was content to soothe my intellectual conscience with such expedients; but in truth they altogether transgress against inward probability and all the psychological laws of historical composition. The more clearly we see that the trial of St. Paul, and above all his appeal to Caesar, is the chief subject of the last quarter of Acts, the more hopeless does it appear that we can explain why the narrative breaks off as it does, otherwise than by assuming that the trial had actually not yet reached its close. It is no use

to struggle against this conclusion. If St. Luke, in the year 80, 90, or 100, wrote thus he was not simply a blundering but an absolutely incomprehensible historian![25]

Mainstream biblical scholarship still dates the writing of the Gospels to around 70 and later. I don't understand why. To me, the evidence for post-70 composition seems wispy. The evidence for pre-70 composition seems extensive and plausible.

CHAPTER 10

The Historicity of the Gospels

I had come to believe that the Gospels were likely based on eyewitness accounts, but such accounts are not necessarily accurate. Any cop or criminal attorney can tell you stories of eyewitnesses getting it wrong (I've been guilty of it myself).[1] Were the Gospels likely to have been *accurate* eyewitness accounts?

Assessing the accuracy of the Gospel accounts is complicated because so much of the material involves supernatural events. On the other hand, I have been surprised at how often suspicious stories check out.

My first surprise came in 2007 when, two years after I read *Mere Christianity*, I was invited to speak on an article I had written for *Commentary*, "Jewish Genius," at the annual Herzliya Conference held on the Israeli coast just north of Tel Aviv. Catherine and I made a holiday of it, renting a car and spending a week driving from Tel Aviv through the Negev to Masada, north to the Roman ruins of Scythopolis, into the Golan Heights, and then to Jerusalem for a few days before driving to Herzliya.

On the way from the Golan Heights to Jerusalem,

we stopped at the archaeological site of Capernaum, the town where Jesus is said to have begun his ministry. There we visited the Church of St. Peter's House, so named because, the guidebook told us, part of the church is suspended over the site of Peter's Capernaum house—the place where the crowd around the house was so dense that a paralytic who hoped to be healed by Jesus had to be lowered through the roof by his friends (Mark 2:1–12).

I walked into the church wholly skeptical, just as I continue to be wholly skeptical of splinters that are said to be pieces of the True Cross. Inside, Catherine and I sat in a pew, looking down at the outline of the foundation of Peter's purported house through a large plexiglass panel. It was as small as the outlines of other houses, but, unlike any of the others, it was octagonal.

Archaeologists have determined that the octagonal foundation dates to the fifth century, when it was part of a larger octagonal church built around it. Excavations under the space demarcated by the small octagon revealed that it had originally been a single room in a house that had undergone an unusual renovation in the second half of the first century, with plastered walls (a rarity at the time), utensils indicating that it was being used for communal gatherings, and graffiti scratched into the walls invoking Jesus Christ. When the larger church was built four centuries later, the small interior octagonal structure was located directly on top of the original room. The evidence strongly suggests that the venerated room is indeed the site of Peter's house.

The Historicity of the Gospels

I walked away from the Church of St. Peter's house bemused. It was the first of many such experiences I've had over the twenty years since I encountered *Mere Christianity*. Repeatedly, an improbable aspect of the Gospels' historicity has held up surprisingly well.

This chapter covers five topics: The accuracy of the Gospels regarding facts that can be checked, "undesigned coincidences" suggesting that different accounts of events in Gospels dovetail, reasons for thinking Jesus's teachings were transmitted accurately, what to make of the miracles in general, and what to make of the resurrection specifically.

Getting the Facts Right

The Gospels are full of the names of places and people and descriptions of Palestinian geography, physical landmarks, and local customs. These lend themselves to fact-checking. To see why evidence that the authors of the Gospels had local knowledge is useful, imagine being a Christian living in Rome during the second century and augmenting the circulating versions of Jesus's life with your own embellishments. What errors or omissions could reveal that you are making stuff up?

Geography

The Gospels name twenty-six towns and seven landmarks in Judea and Galilee. Some of the towns are obscure villages such as Chorazin, Bethany, Bethphage,

Aenon, Cana, Ephraim, Salim, and Sychar.[2] They existed, though they were not marked on any map or mentioned in any writings that someone elsewhere in the Roman Empire might have consulted.

The body of water at the center of Galilee is a lake just thirteen miles in length. No one not from Galilee would have confused it with a sea. Luke, who by tradition is thought to have been a Gentile from Antioch, appropriately calls it a lake. But Matthew, John, and Mark all refer to it as the Sea of Galilee—the terminology in local use.

All four Gospels display specific knowledge about the geography of the region. Matthew and Mark knew that hill country adjoins the Sea of Galilee. Matthew, Mark, and Luke knew of the Judean desert near the Jordan. Luke and Matthew grouped the villages of Chorazin, Bethsaida, and Capernaum—and they were in fact within five miles of each other. Luke and John both knew two different routes from Galilee to Jerusalem. The route through Samaria is referred to in Luke 9 and John 4, and both Evangelists also accurately describe Jesus's approach to Jerusalem ahead of the crucifixion via another route, through the village of Bethany. Matthew and Mark put Jesus in a garden called Gethsemane the night before the crucifixion. Luke locates the same events on the Mount of Olives. No Gospel mentions the fact that "gethsemane" means a press for olive oil.

Peter Williams draws three significant conclusions:

- The writers either knew Judea and Galilee themselves or accurately recorded what was reported by others who were acquainted with the land.
- The information the writers had is consistent with what we would expect if the Gospels were written by their traditional authors.
- The Gospels are not what we would expect from people who made up stories at a geographical distance.

For a comparison group, consider the Gospels of Thomas, Philip, and Judas that the early Church Fathers deemed apocryphal. The Gospel of Philip mentions just three locations—Jerusalem, Nazareth, and the Jordan River. The Gospel of Thomas mentions one—Judea. The Gospel of Judas mentions none.

Palestinian Jewish Names

In 2002, Israeli scholar Tal Ilan compiled and published a daunting feat of scholarship, *Lexicon of Jewish Names in Late Antiquity: Part I: Palestine 330 BCE–200 CE*. She drew on the works of Josephus, the New Testament, ossuary inscriptions, texts recovered from the Judean desert and Masada, and the earliest rabbinic sources. In all, the *Lexicon* contained the names of about three thousand people, enabling Richard Bauckham to assemble frequency distributions that revealed the relative popularity of names in first-century Palestine. Then Bauckham drew on compilations of the names of Jews in Egypt, demonstrating that the frequency of names in the New Testament closely corresponded with the most popular Jewish names in Palestine but not in Egypt. Peter Williams adds information about differences among common Jewish names in Asia Minor and Rome compared to Palestine. "In other words," Williams concludes, "someone living in another part of the Roman Empire would not simply be able to think of Jewish names familiar to him and put them into a story, resulting in a plausible group of names for Palestinian Jews."[3]

Bauckham further noted the phenomenon known as disambiguation of names. In an era without surnames, writers needed to add information to distinguish people with common names. For example, the names *Simon* and *Mary* were the most common male and female names among Palestinian Jews. The New Testament routinely disambiguates them (e.g., Simon *Peter*, Simon *the Cyrenian*, Mary *Magdalene*, Mary *the mother of James and Joseph*). Williams uses the list of the Twelve in Matthew to illustrate. The frequency ranks of their names range from first (the two Simons) to greater than ninety-ninth (Andrew and Thomas). Disambiguators are used for the eight apostles whose names were ranked among the eleven most popular. Disambiguators are not used for the four apostles with names that were uncommon (ranked thirty-nine to ninety-nine and higher). "So not only are the names authentically Palestinian," Williams concludes, "but the disambiguation patterns are such as would be necessary *in Palestine, but not elsewhere*. From this we may conclude that, wherever Matthew's Gospel was written, this list itself most likely took something close to its current form in Palestine" (emphasis in the original).[4]

Once again, the apocryphal Gospels can be used as a comparison group. The Gospels of Thomas and Mary name only seven and five people respectively, limited to the most central characters, compared to the eighty people named in the four canonical Gospels. Judas's Gospel has just two Palestinian Jewish names—Judas and Jesus. The rest are names from the Old Testament and from contemporary mysticism (e.g., Adam, Adonaios, Barbelo, Nebro, Yobel) that are nowhere on the list of Palestinian Jewish names at the time of Jesus.[5]

Undesigned Coincidences

The phrase "undesigned coincidence" refers to a connection that couldn't easily have been planned between two or more accounts of an event which nevertheless fit together like pieces of a puzzle.[6] The undesigned coincidences that can be identified in the Gospels suggest that the accounts are from people who were close to the events and were attempting to relate the facts accurately.

Undesigned coincidences correspond to what happens in real life when different eyewitnesses describe an event. If the witnesses are concentrating on reporting what they recall rather than constructing a story, they will often give a detail that seems inexplicable on its own but makes sense when put alongside the testimony of another witness who reported another, complementary detail. Thus, for example, Mark (3:17) mentions parenthetically that Jesus nicknamed the brothers James and John "Sons of Thunder"—an odd choice for a nickname. But not so odd when you read Luke (9:52–56), which records an incident in which a village of Samaritans refuses to receive Jesus. James and John ask Jesus, "Lord, do you want us to command fire to come down from heaven and consume them?" Mark records that Jesus gave them a nickname (I like to think of Jesus laughing when he did it), and Luke coincidentally gives the reason for it.

Or consider the feeding of the five thousand. Jesus sees the crowd gathering and asks Philip, "Where are we to buy bread, so that these people may eat?" (John 6:5). Why ask Philip, one of the otherwise infrequently mentioned apostles? An explanation comes from an earlier, unconnected verse in John which says that Philip was from Bethsaida (John 1:44). John does not say where the feeding of the five thousand took place, but Luke does: Bethsaida (Luke 9:10). Put the two details together, and it sounds as if Jesus is in effect saying, "You're from around here, Philip. Where do you go to buy bread?"

Or there's my own favorite example of an undesigned coincidence: the portrayals of the personalities of Mary and Martha, the sisters of Lazarus, in Luke 10:38–42 and John 11:17–46 (supplemented with John 12:1–8). I leave it to you as an exercise.

The value of these undesigned coincidences is not that any one of them is conclusive about anything.

But they happen frequently, and the cumulative effect needs to be taken seriously. It is unlikely that such subtle connections could have survived repeated additions to and redactions from the original oral traditions. On the contrary, the simplest explanation is that we are reading separate bits of information that fit together only because the witnesses were artlessly giving their independent recollections of events.

The Accounts of Jesus's Teachings

What about the accuracy of Jesus's teachings as reported in the Gospels? The Gospels' authors could have gotten Palestinian geography and Jewish names right, and the Gospels could show signs that they were reporting eyewitness accounts of the same event from different perspectives—and we could still have a serious problem with the transmission of Jesus's teachings. The Gospels present long passages that purport to be direct quotations of Jesus's words. Is it plausible that they are even close to accurate?

Richard Bauckham's *Jesus and the Eyewitnesses* has a detailed discussion of the literature on ancient Middle Eastern practices (including Jewish ones) of controlled, formal transmission of tradition. Among other things, those practices distinguished between the *discussion* of traditional stories, parables, and sayings and the *recitation* of that material, which was restricted to people who were deemed qualified to pass on the tradition word for word.[7] The disciples, especially the Twelve, constituted that subset. What can we expect of the accuracy of their accounts?

Jesus had been the teacher; they, his students. In an era when oral transmission was the standard way of conveying information from one generation to the next, the task of the student was to memorize the teacher's teachings and recite them verbatim. That was true of all teachers and students. In the case of Jesus, the resurrection experience confirmed his students in their belief that he had been not just a teacher, but The Teacher, and that their responsibility to relay his teachings accurately was momentous.

It's also important to keep in mind that the disciples who traveled with Jesus listened to his core teachings repeatedly. The disciples probably heard Jesus preach the beatitudes and the parables to his audiences dozens if not hundreds of times. Memorization of his words was not just possible but close to unavoidable. Jesus also sent the disciples out on their own evangelizing missions, instructing them to proclaim the kingdom of God and presumably adding instructions about how to go about it (Luke 9:1–2).

These are some of the reasons for thinking that Jesus's teachings as reported in the Gospels are likely to be attributable to him and in some cases are probably close to verbatim. The same considerations apply to the early transcribers of the original texts of the Gospels. Many were professional scribes who were trained and paid to copy exactly what they had in front of them. It seems unlikely that devout transcribers who thought they were transcribing the words of the Son of God felt free to add their own thoughts.

Finally, we can compare how much was changed after about the third century, though we cannot prove what was changed before then (absent new discoveries of ancient scrolls). Peter Williams devotes a long chapter to this issue in *Can We Trust the Gospels?*, using Desiderius Erasmus's first printed edition of the New Testament in Greek in 1516 as a benchmark. Erasmus had to rely on just two existing manuscripts, both dating to the twelfth century. We now have access to manuscripts for all four Gospels dating back to about 350—a thousand years earlier than the manuscripts Erasmus had to rely on. The differences between Erasmus's Gospels and the ones scholars now accept based on their much greater proximity to ancient versions are trivial.[8]

The Miracles

It is in the nature of miracles that the events themselves are inexplicable by either science or everyday experience. That hasn't stopped biblical scholars from discussing whether it is theologically reasonable to believe that God might interfere with the laws of nature, which doesn't constitute evidence one way or the other.

> **SOURCES DISCUSSING MIRACLES**
>
> Several of the sources I listed in the previous chapter discuss the miracles. Some of the best of those discussions are Craig Blomberg's *The Historical Reliability of the Gospels*, chapter 3; Peter J. Williams's *Can We Trust the Gospels?*, chapter 8; and Brant Pitre's *The Case for Jesus*, chapters 9 and 10. Other sources are C. S. Lewis's *Miracles: A Preliminary Study* (1947) and Francis Collins's *The Language of God* (2006), chapter 2. Douglas Geivett and Gary R. Habermas (eds.), *In Defense of Miracles: A Comprehensive Case for God's Action in History* (1997) is a compilation of articles by various biblical scholars (and one by a spokesman for the negative, David Hume's famous essay "Of Miracles").

The miracles demarcate the dividing line between what I will call Catherine's Christianity and Evangelical Christianity. Catherine's Christianity is focused on assimilating Jesus's wisdom and example into her understanding of the universe, the human condition, and how to respond in living her life. She doesn't exclude the possibility that he performed miracles. (What is a miracle, she asks, but another aspect of the Mystery?). But whether he did or didn't perform miracles makes no difference to the practice and value of her Christianity. For Evangelical Christians, the miracles are an integral part of the theology of Jesus's divinity, and Jesus's divinity is in turn central to his role as the Christ and his mission as redeemer.

While I admire Catherine's approach and aspire to emulate it, I confess to more curiosity about the miracles than she has. My bedrock belief is that something needs explaining. This is most emphatically true of the resurrection, which I will discuss separately, but it's also true of the healings and the one-off miracles such as converting water to wine, walking on water, calming a storm, raising Lazarus from the dead, and feeding the five thousand with a few loaves and fishes.

The Christmas Stories in Matthew and Luke

For me, the Christmas stories are beautiful fables. The Murray family has been reading Luke's account aloud every Christmas Eve since Anna was a toddler. But the Christmas narratives in both Luke and Matthew read like folklore and have echoes in other Mideastern traditions. The only plausible eyewitness source is Mary (Joseph is wholly absent from the New Testament after Jesus's adolescence). Mary might have supplied an accurate location of Jesus's birth—the cave near Bethlehem where the Church of the Nativity was built apparently became important to Jesus's followers very early. But when it comes to the virgin birth, the angels appearing to the shepherds, the magi, and Herod's massacre of the children, I have to either believe miraculous events reported by a single observer who had a personal interest in them, which is hard to do, or conclude that someone, perhaps Matthew himself or perhaps the author of the hypothesized Q source, decided that such a great man as Jesus must have had a glorious birth, and that by the time Luke was gathering his source materials in the 50s, the stories had been elaborated.

Explaining Jesus's Success Without Resorting to Miracles
That leaves me with the task of deciding what I think about the miracles that are reported in the Gospels as happening when Jesus's disciples were with him. Jesus was an itinerant Jewish teacher who came out of nowhere, attracted huge crowds wherever he went, and by the time of Palm Sunday was seen as serious threat to the Jewish establishment. If I am not prepared to accept the miracles, I must explain how he did it.

It's not hard. He could have been an exceptional human being who exercised human qualities at an extremely high level. Think of it this way: A man who is 6′ 9″ or taller is four standard deviations above the American male mean, which translates into just one out of roughly thirty thousand American adult males.[9] He is extremely tall but far from unique.

Now consider the healings that Jesus is said to have performed routinely throughout his ministry. They are an obvious and plausible explanation for his rapid growth in fame and attention. Then consider that some people are healers in ways that go beyond technical skills—a phenomenon that physicians themselves recognize in rare colleagues and that has been observed in many cultures for centuries. Some of these gifted people are the equivalent of 6′9″ as healers of both mental and physical ailments. Suppose Jesus was one of them. The accounts of Jesus's healings could be largely true even if the miraculous nature of the healings was exaggerated in the retelling.

Continuing along this line, imagine that Jesus was

also the equivalent of 6'9" in wisdom, fortitude, empathy, sympathy, and charisma. Combine all these qualities, and you are faced with an extraordinary and compellingly magnetic figure, surely unique in all of human history. To paraphrase Arthur C. Clarke, a human that far above the mean in all those human characteristics would be indistinguishable from the Son of God—or, at the least, could easily be mistaken for the Son of God.[10]

Taking the One-Off Miracles Seriously

Still, I am uncomfortable leaving the miracles at that. I am persuaded that the text of Mark represents a reasonably accurate transcription of Peter's memories. Mark's descriptions of Jesus calming a storm, walking on water, and twice feeding a crowd of thousands on a few loaves and fishes are explicit. John, who is otherwise so independent of the Synoptic Gospels, describes the first feeding of the five thousand and mentions the walking-on-water miracle in the same context as Mark (occurring the same night). If I am to treat the Gospels as generally trustworthy, why should I reject the Gospels' four consistent and detailed accounts of the same event? Faith that miracles are not possible? It is a question that I still have not resolved to my own satisfaction.

If the one-off miracles I listed were the only ones reported in the Gospels, I could ignore the internal tension in my position. It's not a big deal. But the Gospels tell of one other miracle, and it poses endless difficulties: the resurrection.

The Resurrection

Why not just assume that the resurrection is too outlandish to be true, and dismiss it as easily as I dismiss the story of the three magi following a star to Bethlehem? Partly because the physical resurrection of Jesus is at the center of orthodox Christian theology, partly because the implications would be so momentous, and partly because the evidence associated with it cannot be easily dismissed.

The Baseline

Let's strip the resurrection story down to statements that are historically secure, independently of specific details in the Gospel accounts:

- Jesus of Nazareth was a historical figure who was crucified in Jerusalem in 30 or 33 CE.
- Jesus's apostles were unsophisticated Palestinian Jews.
- Death by crucifixion was not just a gruesome and painful death for Jesus but a humiliation for everyone associated with him.
- Either the Sanhedrin or the Roman authorities wanted to crush the Jesus movement, had caused the execution of its leader, and might be expected to persecute his followers.
- Within a few decades, Christianity had become a full-blown religion and had spread throughout the Middle East and reached Greece and Rome.

That last statement is a non sequitur. Everything we know about the situation at dawn on Easter morning leads us to assume that the disciples were demoralized and fearful. That's probably a mild way of putting it. Their world had been suddenly and brutally torn apart and their hopes shattered. And yet these same men soon became the world's most effective evangelizers, creating a successful religion with stunning speed.

The implication is that something transformative happened to the apostles and Jesus's other followers after the crucifixion. Coming up with a description of *something* that doesn't look like the resurrection turns out to be harder than you might think. Some of the most important considerations:

The necessity of an energizing event. It seems unlikely that the demoralized and frightened disciples somehow slowly recovered from the trauma of the crucifixion and eventually began to preach. Without an energizing event, the passage of time is more likely to have produced disillusionment. Personal encounters with a living Jesus would unquestionably qualify as energizing events.

Message. For the movement to have acquired momentum, the apostles had to have a galvanizing message. Preaching Jesus's beatitudes and parables doesn't fit that description. A message saying that the Son of God was resurrected does.

Timeline for the origination of the message. There is good reason to think that the galvanizing message was proclaimed very soon. The Pauline letters provide a timeline that is independent of the Gospels.

Alone among the books of the New Testament, the seven undisputed Pauline letters were written within a known window of time by a specific person who includes biographical material about his contact with the Christian movement from its earliest days.[11] Paul reports his role in persecuting an active and growing church prior to his conversion, suggesting that a Jerusalem-based apostolate was functioning within a year or so of Jesus's execution. Believers were to be found elsewhere in the Middle East by the time Paul had his road-to-Damascus conversion (that's why he was going to Damascus), one to three years after the crucifixion.

Furthermore, the Pauline letters indicate that Jesus was already being portrayed as the resurrected Son of God in the earliest years of the church. Paul's first surviving letter is believed to be 1 Thessalonians, written about 49–51 CE. In it, Paul is already writing about a divine and resurrected Jesus and treats this as something his readers already know about. In 1 Corinthians and elsewhere, Paul explicitly associates his teachings with those of the Jerusalem apostles who were eyewitnesses.[12] He had been exposed to those apostolic teachings early. In Galatians, Paul reports that he spent fifteen days with Peter in Jerusalem three years after his conversion, which would date the visit to the latter half

> **WHY SOME OBSCURE PEOPLE IN THE GOSPELS ARE NAMED AND OTHERS AREN'T**
>
> Of the two disciples on the road to Emmaus on the first Easter, one (Cleopas) is named and the other is not. Richard Bauckham has an intriguing hypothesis about this and other cases in which obscure people in the Gospel are named. "I want to suggest the possibility that many of these named characters were eyewitnesses who not only originated the traditions to which their names are attached but also continued to tell these stories as authoritative guarantors of their traditions."[16] His hypothesis is a useful reminder that the accounts being disseminated orally in the early decades of the church were not originally for readers (much less, for readers of print versions millennia later like us), but were memories recounted face to face by people who had been participants in the Easter drama themselves or were invoking the testimony of known participants who were still alive.

The Mystery of Easter

Until now, I have been presenting material that does not rely on the Gospels or Acts. My point has been that *something* transformative happened to the apostles and Jesus's other followers soon after the crucifixion and that *something* enabled the apostles to convince many people quite quickly. Now let's bring Acts and the Gospels into the picture.

The Acts of the Apostles gives us a specific interval within which the transformative event had its effect: Peter first publicly preached a resurrected Jesus on the Jewish day of Pentecost, which falls fifty days after Passover.

The Gospels describe the disciples' demoralization and fear that I argued was inevitable. Peter's triple denial that he knew Jesus is the most dramatic illustration

of the 30s. As Richard Bauckham observes, "t
of conversation."[13] Or, as another New Testam
ar put it, "We may presume they did not spe
time talking about the weather."[14]

SOURCES FOR AN EARLY CHRISTOLOGY

The revisionist position is that the resurrection and the divini
were late inventions, not beliefs of the early church. I canno
that position can be squared with the timeline that can be
from the undisputed Pauline letters. If you are rightly dub
how much trust you can put in my biblical erudition, see Joh
The Birth of Christianity: The First Twenty Years (2005), L
do's *One God, One Lord: Early Christian Devotion and Anc
Monotheism* (2015), and Part 2 (190 pages long) of N. T. W
Resurrection of the Son of God (2003) for detailed evidence
early 50s Paul was already preaching a Christology to whicl
tributed but that was based on fundamentals described by tl
from the beginning.

The availability of credible evidence. An extr
claim cannot be conveyed persuasively w
traordinary evidence. A plausible candidate
timony of many people that they had pers
countered the resurrected Jesus.[15] During th
40s, most of those eyewitnesses were still
living relatively near each other—many in J
with others concentrated in Palestine and t
diately adjacent Roman provinces. Those
spreading the news of Jesus's resurrection co
you don't believe me, just ask . . . ," listing oth
nesses within easy reach.

of this (Mark 14:66–72), along with John's statement that the disciples made sure their room was locked "for fear of the Jews" (John 20:19).

SOURCES FOR THE RESURRECTION

For defenses of a physical resurrection, two magisterial works are N. T. Wright's *The Resurrection of the Son of God* (2003) and Michael R. Licona's *The Resurrection of Jesus: A New Historiographical Approach* (2010). Two more accessible works are Carl E. Olson's *Did Jesus Really Rise from the Dead? Questions and Answers About the Life, Death, and Resurrection of Jesus* (2016), and Dale C. Allison, Jr.'s *The Resurrection of Jesus: Apologetics, Criticism, History* (2021). Two volumes of a projected four-volume work are Gary R. Habermas's *On the Resurrection: Evidences* (2024) and *On the Resurrection: Refutations* (2024).

What scenarios could explain away the New Testament accounts of Easter morning and its aftermath?

A scenario that says the disciples concocted the whole story is hard to accept. First, remember who the apostles were. They were not advisors, partners, collaborators, or public relations strategists. They were not historians, poets, or novelists. Instead, probably all of them but Matthew were illiterate manual laborers, unsophisticated followers of an itinerant Palestinian preacher. Now consider how extraordinary, unprecedented, and revolutionary their concocted story was. As N. T. Wright documents in excruciating detail, the resurrection was not a repackaged version of anything that had come before.[17] Jesus's apostles had nothing in Jewish tradition that could have inspired them. For that matter, neither was there any pagan tradition that they might have drawn upon. Jesus's apostles were unique in asserting that a specific person at a specific historical time, witnessed by others, had been bodily returned to life and that his resurrection signaled the beginning of God's new creation.

A scenario that says the disciples lied about the tomb being empty is hard to sustain because the location of the tomb was known to the Roman and Jewish authorities and to the general population. When Peter started preaching the resurrected Jesus, the authorities could have discredited him by opening the tomb and exhibiting the moldering body. If that had happened, the movement would have collapsed before it got started. Apparently, neither the Roman authorities nor the Jewish Sanhedrin denied that the tomb was empty.

A scenario that has the disciples stealing the body and hiding it requires one to believe that the tomb was unblocked and unguarded, that the frightened and demoralized disciples put together a plan to steal the body during the hours immediately after the trauma of the crucifixion, and that they executed that plan so adeptly that it was never exposed. This unlikely scenario also requires me to believe that the disciples spent the rest of their lives evangelizing for Jesus despite suffering continual hardships, eventually going to their deaths still insisting that Jesus had been resurrected, even under torture, while knowing all along that they were preaching a lie.

The obvious reasons to dismiss the resurrection run up against logical implausibilities. Trying to summarize the more complicated issues would push us into a maze of controversies that would take dozens of pages even to outline. I'll limit myself to this: If you're trying to decide what you make of the historicity of Christianity, you have no choice but to read into the literature on the resurrection. The good news is that the exercise is endlessly fascinating. I recommend it. But be warned: Sooner or later you will have no choice but to grapple with the bizarre story of the Shroud of Turin.

The Shroud of Turin
Catherine has been rolling her eyes ever since I first exhibited an interest in the Shroud of Turin. It's one more instance of my "beside the point" absorption in questions of historicity.

You may be rolling your eyes for another reason—the shroud has been proven to be a medieval fake, right? No, actually, the Shroud of Turin is one of the most exhaustively studied objects in the world and continues to be one of the most baffling.

Woven of linen, 3.7 feet wide and 14.5 feet long, the shroud shows faint and indistinct frontal and dorsal images of a human male. It is purportedly the cloth used to wrap Jesus's body after the crucifixion. Early church tradition puts the shroud in Edessa and later in Constantinople during the first millennium CE. The documented history for the Shroud of Turin goes back to the mid-fourteenth century, when it showed up in Lirey, France. Since 1694, it has been stored in a chapel adjacent to the Turin Cathedral.

Taking Religion Seriously

The first photograph of the shroud was not taken until 1898. It created a sensation. The developed photograph was unexceptional, capturing the same faint outlines that people had been viewing for the previous seven centuries. In vivid contrast, the *negative* of the photograph revealed detailed frontal and dorsal images of an adult male who had been beaten and scourged, with numerous puncture wounds on the top and back of his scalp and forehead, spike-sized holes in both wrists and feet, and a wound in the right side of the chest.

As early as 1902, it was suggested that the image densities in the negative appeared to vary inversely with the distance between a cloth draped over a human figure and the flesh of that figure—darkest where the cloth was in closest contact (e.g., the tip of the nose), lighter where the distance was greater (e.g., the neck). In 1978, this hypothesis was verified when images of the shroud were analyzed by a VP-8 Image Analyzer, a machine designed to convert image brightness into a three-dimensional representation. When applied to normal photographs, a VP-8 produces an uninterpretable jumble of light and dark shapes. When applied to the shroud it produced a correctly proportioned display of the image seen in the photographic negative. The images on the shroud encode three-dimensional information.

The VP-8 analysis was part of the most extensive scientific examination of the shroud. It was called the Shroud of Turin Research Project (STURP). The research team consisted of thirty-three American scientists drawn from the Los Alamos National Laboratory, Sandia National Laboratories, and the faculties of physical science departments of major universities and other specialized research organizations. STURP's primary objective was to determine how the image was created.

Explaining how a medieval forger could have encoded three-dimensional information was daunting, but at least it left open hypotheses that involved a cloth draped over a human figure. Explaining the results of the chemical analyses was more than daunting. The findings are literally inexplicable for three reasons.

First, no paint or pigment was found on the linen fibers of the shroud. Some of the faint stains on the shroud have been conclusively identified as human blood. None of the stains was produced by fluids associated with a decomposing corpse.

Second, analyses at the Jet Propulsion Laboratory using a microdensitometer revealed that the coloration on the shroud is microscopically directionless. Any known method of creating an image by hand (e.g., with a brush) has two-dimensional directionality.

Third, the straw-colored image on the shroud is bafflingly superficial. Each linen thread in the shroud is composed of smaller fibers ten to twenty times thinner than a human hair. The color on the shroud is always on the top two or three fibers in the thread—it never penetrates deeper than two microns (millionths of a meter). On a colored fiber with a diameter of fifteen microns, the cellulose within is uncolored. Colored fibers are side by side with uncolored ones.

The implications are that any non-pigment coloring agent (e.g., an acid) that had been applied to the surface of a fiber must cover the entire circumference of the fiber and yet be confined to two microns of depth, leaving the interior uncolored, and have no effect on adjacent uncolored fibers. Doing this would require a brush with one bristle no bigger than a hundredth of a millimeter in diameter—and access to a high-power microscope. These findings ruled out all known artistic ways of creating the image.[18]

The STURP team then explored a variety of hypotheses involving other mechanisms for creating the image—for example, the "scorch" theory that a statue of a human had been heated and the linen placed over the hot metal surface. All the alternatives had insurmountable problems. "Briefly stated," the STURP technical report concluded, "we seem to know what the image is chemically, but how it got there remains a mystery. The dilemma is not one of choosing from among a variety of likely transfer mechanisms but rather that no technologically credible process has been postulated that satisfies all the characteristics of the existing image."[19] That finding has not been successfully challenged in the forty-three years since the STURP team published its technical report. It constitutes the core mystery of the shroud.

Collateral findings involving the blood flow from the wounds, the scourge marks, and other details of the image are consistent with the biblical account and with anatomical and historical evidence. The material gathered during the 1978 investigation also included samples of dust, pollen, and other organic and inorganic evidence that were lifted from the shroud using adhesive tape. They have subsequently produced evidence that the shroud was once in the Middle East and more specifically in Jerusalem.

The linen, which is of fine quality, has a distinctive weave that was common in antiquity in the Mideast but not in medieval Europe. A complex style of stitching is identical to the one used on an ancient garment found in Masada. A few of the dozens of types of pollen and other plant materials that were lifted from the shroud grow in the places that tradition identifies with the shroud's journey (Edessa, Constantinople, Lirey, and Turin), but a large majority grow near Jerusalem. About half of them are found only in the Mideast, never in Europe. A few are confined specifically to the area adjacent to Jerusalem. The time of blossoming for all the Mideast pollen includes the period during which Passover can occur.

The shroud contains microscopic traces of a rare type of limestone. Analyses using a high-resolution scanning ion microprobe at the University of Chicago's Fermi Institute matched the limestone traces on the shroud with the limestone on the same rock shelf as the Holy Sepulcher and Garden Tomb. It was the only match out of the ten limestone tombs that were examined.

Those are just a few of the many strange features of the shroud that were revealed by the STURP analysis and follow-up studies. Then, in 1988, carbon dating put the creation of the shroud at 1325 plus or minus 65 years—the appropriate range if the image is a forgery. For many, this closed the case. But carbon dating the shroud was controversial even before it was implemented—the danger of a contaminated sample was thought to be high. That danger was apparently realized. Raymond Rogers, the coauthor of the original STURP technical report, studied the sample and concluded that "[t]he combined evidence from chemical kinetics, analytical chemistry, cotton content, and pyrolysis/ms proves that the material from the radiocarbon area of the shroud is significantly different from that of the main cloth. The radiocarbon sample was thus not part of the original cloth and is invalid for determining the age of the shroud."[20]

In 2019, a new test of the shroud was conducted with an independent method of dating: Wide-Angle X-ray Scattering, examining the degree of natural aging of the cellulose that constitutes its linen. It produced an estimated age of nineteen to twenty-one centuries and ruled out the possibility that the numbers produced by the carbon dating are correct. From the technical article making that case: "To make the present result compatible with that of the 1988 radiocarbon test, the TS [Turin Shroud] should have been conserved during its hypothetical seven centuries of life at a secular room temperature very close to the maximum values registered on the earth."[21]

> **SOURCES FOR THE SHROUD OF TURIN**
>
> If you are new to this story and need assurance that it's not based on pseudoscience, I recommend that you begin with the summary of the original STURP report by L. A. Schwalbe and R. N. Rogers, "Physics and Chemistry of the Shroud of Turin: A Summary of the 1978 Investigation," *Analytica Chimica Acta* 135, no. 1 (January 1982):3–49, available at shroud.com. The science in the article is as hardcore as anyone could ask. And if the recent redating of the age of the shroud to the first century seems suspiciously convenient, you may read the technical paper describing the test: Liberato De Caro, et al., "X-ray Dating of a Turin shroud's Linen Sample," *Heritage* 5, no. 2 (April 2022):860–70, also available at shroud.com.
>
> Amazon sells a few dozen books on the shroud, including an entertaining novel, Christopher Buckley's *The Relic Master* (2015), which imagines artist Albrecht Dürer's mad-scientist method for forging the shroud. For a journalistic account of the STURP investigation by a member of the team, see John H. Heller's *Report on the Shroud of Turin* (1983). For recent detailed reviews of the accumulated findings, see *The Shroud of Turin: First Century After Christ* (2020) by Giulio Fanti and Pierandrea Malfi, and Gerard Verschuuren's *A Catholic Scientist Champions the Shroud of Turin* (2021).

Where do I come out? The shroud is so very strange that its creation by a medieval forger would be nearly as incredible as the resurrection. If the redating of the shroud to the first century is confirmed by an independent method or is otherwise reinforced, it will be difficult for me to deny that the shroud wrapped the body of the crucified Jesus. But it is impossible to reconcile the characteristics of the shroud with exposure to a decomposing corpse. I think I would have to accept the shroud as evidence of a physical resurrection, with all that implies. For now, I cautiously associate myself with John Polkinghorne's position on the resurrection in *The Faith of a Physicist*:

> The only explanation which is commensurate with the phenomena is that Jesus rose from the dead in such a fashion (whatever that may be) that it is true to say that he is alive today, glorified and exalted but still continuously related in a mysterious but real way with the historical figure who lived and died in first-century Palestine.[22]

Nothing in this chapter means that I think the Gospels we have today are pristine versions of the original accounts. I accept that many specifics of the revisionists' evidence for augmentation and redaction are correct. But when it comes to the accounts of Jesus's ministry and his teachings, I am persuaded that the Gospels contain a great deal of material from eyewitnesses who were reporting real events, who might reasonably be expected to have memorized Jesus's core teachings, and who may be expected to have repeated them accurately.

    I am writing as someone who grew up in the Christian faith, fell away, and, upon further consideration, has returned to it. If instead you grew up in a secular setting and have never given Christianity much thought, I hope that one takeaway from my experience is that it would be worth your while to do so.

CHAPTER 11

What's the Point?

I set out to demonstrate that religion is something that nonbelievers can and should take seriously. I have used my own experience to illustrate that you can. I still haven't spelled out *why* you should take religion seriously.

One answer is personal—what I've taken away from the process. I'll get to that, but I will begin with two answers that apply more generally. The first is that we live at a time when the relationship of religion to science has shifted in ways that require nonbelievers to reconsider their position. The second is that C. S. Lewis's position on the Moral Law generalizes to a broader issue, one that nonbelievers should try to resolve for themselves.

GOD OF THE GAPS REVISITED

"God of the Gaps" refers to the tendency of people to confuse gaps in scientific knowledge with evidence for the existence of God. When modern science took off early in the sixteenth century, just about everything was a gap. Over the next four centuries, science filled in most of them. Science displaced Earth from the center of the cosmos. Science discovered that God has nothing to do with earthquakes, lightning, and other mystifying natural phenomena. When, at the beginning of the nineteenth century, Napoleon asked Pierre-Simon Laplace why his explanation of the solar system in *Mécanique Céleste* did not mention God, Laplace could answer, "I had no need of that hypothesis." It was a statement of fact.

By the end of the nineteenth century, science had delivered more body blows to the God of the Judeo-Christian Bible. Charles Lyell replaced a biblical Earth that was less than six thousand years old with a geological Earth that was millions (later billions) of years old. In the popular imagination, Charles Darwin replaced Adam and Eve with chimpanzees. Freud replaced sin with superego deficiency.

Science also made God less important as a resource for coping with the challenges of life, especially over the course of the twentieth century and especially in the West. Wealth, security, technology, and medical science have left the fortunate among us free to forego prayer. We apparently have no need of God's help to live comfortable and entertaining lives.

And yet even as life without God was becoming easier in the twentieth century, the relationship of science to religion was taking a strange turn. Instead of filling gaps, scientists began to discover new gaps for which science did not have an explanation, but religion did.

This altered relationship between science and religion hinged upon how the universe works and how consciousness works.

How the Universe Works
The opening year of the twentieth century saw the publication of Planck's Law, introducing what would become known as quantum mechanics—which in turn taught us that the universe is not only stranger than we knew, but stranger than we could imagine. Quantum mechanics changed physics from Newton's tidy clockwork universe to a wonderland of seemingly contradictory dynamics. Particles behave like waves and particles at the same time. The act of observing a particle affects its state. A change in the state of one particle changes the state of a distant "entangled" particle instantaneously, faster than the speed of light.

The mathematics of quantum mechanics were amazingly precise, but the underlying reality of what the math described remained mysterious and counterintuitive. As Richard Feynman said in a lecture at MIT more than sixty years after Planck opened the field, "I think I can safely say that nobody understands quantum mechanics." He continued:

> So do not take the lecture too seriously, feeling that you really have to understand in terms of some model what I am going to describe, but just relax and enjoy it. I am going to tell you what nature behaves like. If you will simply admit that maybe she does behave like this, you will find her a delightful, entrancing thing. Do not keep saying

to yourself, if you can possibly avoid it, "But how can it be like that?" because you will go down the drain, into a blind alley from which nobody has escaped. Nobody knows how it can be like that.[1]

No one since has had the chutzpah to contradict Feynman. For many physicists, the sensible thing to do has been to just "shut up and calculate," as physicist David Mermin put it.[2] In a universe run via quantum mechanics, the God hypothesis did not seem so outlandish.

The development of quantum mechanics was just the first of several paradigm shifts during the twentieth century.[3] The 1920s saw revolutionary discoveries about the size and composition of the universe. In the 1960s, the Big Bang theory of the universe finally achieved acceptance—the "let there be light" origin story of Genesis expressed in equations.

These shifts were followed in the 1970s by the discovery of the brute facts of the Big Bang, which revealed that the odds against a universe that permits life are trillions to one. Science had discovered a new gap. Science could provide no non-religious explanation without hypothesizing that our universe is one of millions. Religion had a simple answer: God had intentionally created a universe that permits life.

As I wrote in chapter 4, my reaction is to accept the religious answer. Accepting it does not require belief in a personal God. Perhaps the existence of life is a by-product of God's intention and we have no idea what that intention might be. But accepting that we live in an intentional universe is tantamount to acknowledging the existence of a God for whom the concept of *intention* applies. That's a huge departure from both atheism and agnosticism.

That may not be your reaction. But if not, why not?

How Consciousness Works

In today's world, the strict materialistic view of consciousness is roughly in the same fix that Newtonian physics was in 1887, when the Michelson-Morley experiment proved that the speed of light doesn't behave as Newton's laws said it should. It took eighteen years before Einstein's theory of special relativity provided a model that accommodated the anomaly. Now the evidence about certain types of terminal lucidity and near-death experience has introduced a powerful challenge to the materialist position on consciousness. They amount to anomalies that I think eventually must lead to a paradigm shift.

Try to imagine scientific advances that let us explain these phenomena *without* accepting that consciousness can exist independently of the brain. The explanation for terminal lucidity must involve discovery of a completely unknown capability that enables the brain to produce cognitive functioning even when the regions of the brain now thought to produce consciousness are no longer functional. The explanation for the factually accurate out-of-body observations in NDEs must explain the acquisition of visual knowledge without the use of the eyes and of aural information without use of the ear—and in the absence of blood flow, oxygen, and brain-stem reflexes.

I cannot see how that is possible, nor do I think it is possible to successfully debunk all the reported cases of terminal lucidity and out-of-body acquisition of information. Too many well-documented cases are out there.

Resolving the anomalies doesn't require starting from scratch. Existing neuroscience explains much about brain function under ordinary circumstances, just as Newtonian physics explains the movements of terrestrial and celestial objects under ordinary circumstances. Physicists weren't required to discard Newton's laws, but to supplement them with laws that apply to extraordinary objects and circumstances at the macro level (the theory of relativity) and micro level (quantum mechanics). Similarly, neuroscientists are not required to discard their current model of brain function, but instead to start exploring the hypothesis that strange things happen when death is imminent. That won't happen unless mainstream neuroscience begins to take that possibility seriously and approach the evidence from terminal lucidity and NDEs in the spirit that William James championed for the investigation of paranormal phenomena.

At this point, I should confess that I have found myself oddly resistant to accepting the evidence for consciousness independent of the brain despite my dispassionate judgment that the evidence is valid, whereas I did not resist evidence that we live in an intentional universe.

Why? Both beliefs involve things that science cannot explain—miracles, in the ordinary sense of that term. But the creation of the universe is abstract, and I can construe it as removed from anything involving me as an individual, whereas the existence of a consciousness that is independent of brain function is directly and intimately relevant to me. To confront the straightforward implication of the evidence—that I have a soul—is intimidating. Accepting that humans have souls puts on the table every religious belief about what happens when we die, from reincarnation to heaven to nirvana. It requires revisiting every religious doctrine about sin, redemption, and salvation that I formerly assumed I didn't have to think about because I already knew they were ridiculous.

Another and more prosaic explanation of my resistance is the fear of what the other members of my tribe will think. Recall Catherine's way of expressing what she learned about religion when she went to college: "Smart people don't believe that stuff anymore." One of the tribes to which I belong is the tribe of smart people. For me to accept the evidence regarding terminal lucidity and near-death experiences (and to publish!) will lower other members' opinions of me, including the opinions of many people whom I admire and whose good opinion I value. I don't want to be thought credulous and foolish and get kicked out of the tribe.

If you find yourself reluctant to give up strict materialism for similar reasons, try to get over it. It's understandable that you may not be convinced by my highly condensed summaries of the evidence for consciousness independent of the brain, but I have been drawing on a *lot* of empirical material. Bring to the question of your soul the same critical faculties that you have brought to more prosaic empirical questions, using the same tool—good evidence—that you've always used to make up your mind. You don't need to reach "Yes I believe absolutely" decisions. It's okay to think in terms of probabilities. Just don't take for granted that what gets labeled as *supernatural* cannot possibly exist. There's good evidence that it does.

THE BACKDROP TO THE MORAL LAW

In *Mere Christianity*, C. S. Lewis uses the Moral Law as a lead-in for exploring Christianity, and that's how I treated his discussion in chapter 8. But in another book (*The Abolition of Man*), Lewis discusses how the Moral Law raises an issue that cuts across religions and philosophic systems: the doctrine of objective value, "the belief that certain attitudes are really true, and others really false, to the kind of thing the universe is and the kind of things we are."[4] We're no longer talking merely about cross-cultural similarities in ethical precepts, but about a broader conception of the transcendent source of existence.

The centuries on either side of 500 BCE saw a transformation in human understandings of the cosmos. In northern India, the Upanishads had supplied the philosophical underpinnings of Hinduism. In northeastern India, Siddhartha Gautama achieved enlightenment and founded Buddhism. In China, Laozi wrote the *Tao Te Ching*, the founding text of Taoism. In the Middle East, the writings that constitute the Torah were woven together and codified.[5]

In terms of religious practice, these developments could hardly have been more different. Hinduism had a pantheon of gods, Judaism had one, Taoism and Buddhism had none. But at the core of all of them, as of Christianity, was the concept of an eternal force that represents the natural order of the universe, and all of them proclaimed similar principles of human flourishing.

Among these similarities, the most striking to me is the emphasis on altruism—in the Christian tradition, *agape*. For Hinduism, generosity and compassion generate positive karma, which in turn determines your fate in your next life. For Mahayana Buddhism, compassion and loving-kindness should lead those who are on the way to enlightenment to delay their liberation to help others. For Laozi, compassion is the first of his "three treasures." For Judaism, the commandment in Leviticus 19:18 to "love your neighbor as yourself" is to be accompanied by giving to those in need and acts of loving-kindness without expecting anything in return.

I am also struck by the degree to which the principles of human flourishing in all five religions draw upon the "cardinal virtues" first identified in classical Greek philosophy—courage, justice, temperance, and prudence. The vocabularies vary, but all five traditions agree that the underlying qualities of fairness, self-restraint, fortitude, and wisdom are the moral foundation that enables people to exercise generosity and compassion to good ends.

In broad strokes, the two approaches to thinking about these similarities are that they can be explained by evolutionary psychology, or that they can be explained as imperfect human descriptions of underlying transcendental truths that are objective and unchangeable.

If you studied ethics in college, I'm sure you were exposed to the argument common to most college courses on the topic: Ethical systems are human creations. "Morality" is situational and subjective. If you studied evolutionary psychology, you can describe how a wide variety of supposedly "moral" principles have their origins in evolutionary selection pressures.

As I noted in chapter 8, I am a fan of evolutionary psychology and agree with many such arguments. But I also noted that evolutionary psychology faces a difficult challenge in explaining what seems to be an inborn impulse to altruism, and more generally a recognition of "the right thing to do" even when it has no evolutionary utility.

This brings us to the second approach. How much importance should be assigned to the ancient arguments for transcendental truths? Most of you, and certainly I, are not knowledgeable about all the details of those ancient arguments. In my own case, I know a fair amount about Christianity, Judaism, Taoism, and Buddhism; little about Hinduism.

It is worth our while to dig deeper. In my view, the founders of all five religions were apprehending a portion of the objective and eternal truths about "the kind of thing the universe is and the kind of things we are." The different expressions of those truths have been shaped by the eras and cultures in which they were apprehended, but I am confident they have one thing in common: All were produced by people with spiritual insight far beyond mine. If your goal is to decide what you think about objective right and wrong, I suggest that it is obtuse to ignore that cross-cultural body of work.

One more thought: Even if you were born as late as 1990, you have observed in your lifetime tumultuous changes in the secular received wisdom about what is right and wrong, good and evil. I was born in 1943, and have witnessed 180-degree flips in the secular received wisdom on child-rearing, marriage, divorce, euthanasia, abortion, acceptable public behavior, responsibility for the consequences of one's actions, and virtually everything about human sexuality. Many of these changes do not appear to have been for the better. Doesn't the evanescence of moral principles in the present age suggest a special need to seek moral bedrock?

Surprised by Belief

If I am writing a book about taking religion seriously, it seems incumbent on me to state publicly some specific things it has done for me. Those things are idiosyncratic and probably irrelevant to your situation. You may skip them without loss.

The short answer is that taking religion seriously has done far more in retrospect than I realized while it was happening. I have periodically discovered that I was thinking differently about God and life than I had thought some years earlier. I will borrow from the title of C. S. Lewis's autobiography, *Surprised by Joy*. I have been surprised by belief. Some specifics:

An Intentional Universe
I live day to day believing that I am part of an intentional universe, not Richard Dawkins's universe of "no design, no purpose, no evil, no good, nothing but pitiless indifference."[6] That's a big change from the way I lived my life thirty years ago.

Interconnectedness
One of the most common reports about meditation, whether of the Taoist, Buddhist, Hindu, Christian, or other variety, is a heightened perception of the interconnectedness of things. Psychedelic drugs can produce the same kind of awareness. It is also one of the most common reports among NDE experiencers: They perceived a oneness across apparently disparate aspects of their lives and, for that matter, the universe. I've never experienced it personally, but I believe this interconnection to be real. I hope to experience it eventually.

The Moral Law
As I've already revealed, I have found C. S. Lewis's argument for the Moral Law to be exasperatingly sticky.

What's the Point?

Writing this book has made me recognize that I have internalized a gut-level belief that he was right: Our impulses to "do the right thing" are God's way of revealing himself to us, and what God has revealed is the pervasive role of *agape* in living a good human life.[7]

Forgiveness of Sins

I recall sitting in Meeting one First Day in the 1990s thinking about something I had done in my twenties of which I was ashamed. Christianity tells me that I can be forgiven. My attitude was that I shouldn't be forgiven. I should feel awful about it for the rest of my life. Serves me right. Now I believe that I was being incredibly egotistical. Who was I to think I was so all-wise that I ought to overrule God's judgment? It's up to him. My task is to be unreservedly repentant. I can do no more than that. Once again, I am not reporting a process in which I went step by step from position A to position B. Rather, I now recall that episode from thirty years ago and think about how silly I was then. And whereas I still feel awful about my sins—and I've grown more comfortable with using the word *sin*—I also have a sense that I have received forgiveness. God's grace has become real to me.

Death and the Afterlife

A few NDE experiencers report an afterlife of darkness and despair, but the rest report variations on an afterlife filled with light, love, and a deep understanding of how everything fits together.[8] Over the years since

I first became aware of the NDE literature I have gradually assimilated those accounts into my set of beliefs. I know that to be true because for some years I have been advancing into old age untroubled by the prospect of death.

That wasn't always true. In my forties and fifties, I would occasionally confront the reality of my eventual death and feel existential dread. I can't pinpoint when my attitude changed from dread to curiosity, but it did. Don't misunderstand me. I will be shocked and dismayed if a day comes when my physician tells me that I've got three months left. But I will continue to believe that my consciousness may survive death. I believe that an afterlife is likely, though I'm not sure how likely, nor do I have any idea of its contours.

In the process of thinking about death and an afterlife I also happened on two thoughts (I can't remember when or why) that have served me as launch points—third-rate koans, you might say—for thinking about a universe in which God is not cruel but neither does he save us from ourselves.

Maybe hell is moral clarity about bad behavior. I imagine that God does not send anyone to hell. Instead, God gives us perfect moral clarity about our lives when we die, allowing no escape from recognizing how much of the wrong we did was our own doing, not to be explained away by influences beyond our control. If in life we have done great wrong without regret, in death our burden of guilt amounts to hell.

What's the Point?

Maybe souls can commit suicide. This is a variant on Pascal's Wager (if you wrongly believe in God, there's little loss; if you wrongly disbelieve, you face infinite loss). Perhaps the soul survives death only if the person about to die accepts the possibility of survival.

Neither of these thoughts amounts to a belief, but they are sobering. Like the prospect of being hanged in a fortnight, the possibility that they might be true concentrates my mind wonderfully.

Besides my newfound belief that an afterlife is a realistic possibility, I have experienced two other shifts in my feelings about death. First, a strong sense has settled in me that life has a proper trajectory. I'm saddened by those who seem desperate to prolong life and want science to enable indefinite life spans. I cannot imagine wanting that. This life has been wonderful for me, but it wouldn't stay wonderful for much longer than I'm likely to live.

Second, I've come to believe that it is not appropriate to decide when to end my life. Since early in our marriage until a few years ago, I had a private joke with Catherine about her responsibility if I got old and was mentally losing it—she was supposed to know when the right time had come to "put the white powder in my gin." Suicide properly done, with assistance if necessary, seemed to me a good thing. I've changed my mind. Short of suffering unmanageable pain, I no longer think I have the right to interfere. I may not have the gift of faith, but I do believe that life itself is a precious gift and that I should treat it as such.

Reclaiming Access to the New Testament
Thirty years ago, I assumed that the Gospels were mostly folklore, and I didn't pay attention to what was in them. The New Testament is full of wisdom. Studying the authorship, timing, and historicity of the Gospels has restored my access to it.

Reassessing Jesus
My plunge into the historicity of the Gospels was triggered by C. S. Lewis's trilemma. Am I really required to choose among just three alternatives: Jesus was a liar, a lunatic, or the Son of God?

All that reading cut into my wiggle room. I lost the option of believing that claims for Jesus's divine nature were a later invention of the church. I now accept that Jesus of Nazareth represented himself as having a unique relationship with God—and suspect that he had such a relationship in fact.

I also believe that getting much more specific in describing Jesus's relationship to God is impossible, not just for me but for everyone. I say that in part because of my belief that God must not be anthropomorphized, and that to talk about a human-like relationship with God violates that rule. John Polkinghorne, a professor of mathematical physics at Cambridge for eleven years before resigning to study for the priesthood, has an approach to thinking about Jesus's divinity that appeals to me: "I think that the titles assigned to Jesus . . . play the role that models do in scientific investigations. They give useful but limited insight. Because their roles are frankly heuristic and exploratory, they can be used with a considerable degree of tolerance of unresolved difficulties."[9]

The larger problem is one that Christian theologians have struggled with since Christian theology began: how to describe the duality of Jesus's human and divine natures. I have returned to a way of thinking about that conundrum that I learned when I was about twelve years old in the confirmation classes I took from the Reverend Lowell McConnell at the First Presbyterian Church in Newton, Iowa. He asked his small class of pubescents to imagine taking a jar to the seashore and filling it with seawater. He then asked us whether our jar of water was the same as the ocean. We all said no. He said it's the same with Jesus. Jesus wasn't God. He was as much God as you can get into the human jar.

The analogy is so striking and so apt that I assume the Reverend McConnell, a graduate of Columbia's Union Theological Seminary, didn't think it up himself, but I've been unable to find the source (perhaps some reader who knows will tell me). In any case, it's a way of thinking about the relationship of Jesus to God that I can get my mind around.

I recently discovered that I believe in that relationship at a deeper level than I knew. For the last thirty-odd years, I have spent the better part of an hour awake every night, usually around three in the morning. For the first few years, I concentrated on ways of making myself go back to sleep. As time went on, I realized that I was getting some good ideas during the wakefulness, usually about problems associated with whatever book I was working on.

During one such wakefulness a few months before writing these words, I was thinking about what it would be like to meet great religious figures from the past such as Gautama, Laozi, Moses, and Jesus. It would be fascinating, of course, to see what they were like in person, and I would naturally treat all of them with the utmost respect. Unbidden, it came to me that I would treat Jesus differently. With reverence.

I began by warning you that my way of taking religion seriously was more arid than I would prefer, and that remains true as I write. Catherine compares her evolving faith to a light on a rheostat that has gradually gotten brighter over the years. Nothing like that has happened with me. I have yet to experience the joys of faith. When I'm around Catherine and others who have, I sometimes feel like a little boy whose nose is pressed against the window, watching a party he can't attend.

I'm not done trying to join the party. Perhaps the door will open eventually. But even if it doesn't, I have much to be grateful for. My haphazard pilgrimage has already led me to believe that I live in a universe made meaningful by love and grace. That's a lot.

Acknowledgments

Undertaking this project would never have crossed my mind without Nick Eberstadt's suggestion. It turned out to be one of most rewarding—and fun—things I've written. Thanks too to Nick's co-interviewer, Karlyn Bowman, whom I first met in the mid 1980s even before AEI hired me. She has been an inspiriting presence for me ever since.

Thanks go to Pete Wehner for introducing me to *Mere Christianity*, a pivotal event, and, a decade later, for including me in a small book club he organized. Half a dozen of us gathered at Pete's home near Washington every month to discuss a religious book recommended by one of our members. I will not name my companions, but they should understand what an important part of my pilgrimage those meetings were.

Thanks go as well to Roger Kimball for deciding to publish this odd little book, so far removed from my usual material, and to Elizabeth Kantor for her meticulous copy edit of the text.

Goose Creek Meeting has now been a central part of Catherine's life and mine for more than three decades.

Acknowledgments

I have contributed far less to the Meeting than Catherine, but I am no less grateful for the fellowship of our Goose Creek Friends.

Catherine has been my beloved partner since 1981 and my first editor, contributing to the voice and substance of everything I've written. But her role in *Taking Religion Seriously* stands apart. She is the reason I had anything to write about.

Notes

Chapter 2: Perceptual Deficit

1. These other abilities probably have a positive statistical relationship to IQ. I don't have hard numbers, but based on IQ's relationship to a variety of other personality and skill characteristics, IQ is likely to have a correlation in the region of +.3 with the ability to appreciate music and visual art, and with spirituality. But if you look at a scatterplot of two variables with a correlation of +.3 (ask an AI program to draw you one), you can see how easily one can be high in IQ but mediocre or low in those abilities.

Chapter 3: Moving off Dead Center

1. Eugene Wigner, "The Unreasonable Effectiveness of Mathematics in the Natural Sciences," *Communications on Pure and Applied Mathematic* 13, no 1 (February 1960):1–14, https://webhomes.maths.ed.ac.uk/~v1ranick/papers/wigner.pdf.

2   The late Charles Krauthammer and I shared a love of chess. Our love was unrequited—Charles was a stronger player than I, but we were both patzers. We formed an informal chess club in the early 1990s. In the wake of the furor over *The Bell Curve* in 1994, we started calling it the Pariah Chess Club. The club continued to meet every Monday evening at Charles's home in Chevy Chase until the early 2000s, when Charles became a regular on the Fox network's *Special Report*—a sad event in my life. Charles described the club in a column that is included in his best-selling book, *Things that Matter: Three Decades of Passions, Pastimes and Politics* (Crown Forum, 2013).

3   Martin Heidegger originated the specific phrase, "Why is there something rather than nothing?" In Heidegger's German, it is "*Warum ist überhaupt Seiendes und nicht vielmehr Nichts?*" Martin Heidegger, "Was ist Metaphysik?" (lecture, University of Freiburg, Freiburg, Germany, July 24, 1929).

4   Stephen W. Hawking, *A Brief History of Time: From the Big Bang to Black Holes* (Bantam, 1988), 1.

5   Since I am writing from a Christian perspective, I use Christian terminology—in this case, referring to "the Old Testament" rather than "the Hebrew Bible."

Chapter 4: The Brute Facts of the Big Bang

1   Robert Jastrow, *God and the Astronomers* (W.W. Norton, 1992), 107. Quoted in Francis S. Collins, *The Language*

of God: A Scientist Presents Evidence for Belief (Free Press, 2006), 66.

2 My view of this book's purpose, including its autobiographical nature, means that I must try to remain true to my knowledge and frame of mind in earlier years. This raised a problem when Ross Douthat published *Believe: Why Everyone Should Be Religious* (Zondervan, 2025) in the same week that I finished the draft of *Taking Religion Seriously*. I could not appropriately use anything I might learn from *Believe*, and the easiest way to ensure that I didn't was not to read it. I have faced a similar problem with the perceptive comments I have gotten from friends who read the manuscript. I have taken advantage of them only if doing so doesn't make me sound more erudite than I was when my views were evolving.

3 Stephen C. Meyer, *Return of the God Hypothesis: Three Scientific Discoveries that Reveal the Mind Behind the Universe* (HarperOne, 2021), 381.

4 Collins, *The Language of God*, 68.

5 Martin Rees, *Just Six Numbers: The Deep Forces That Shape the Universe* (Weidenfeld & Nicolson, 1999), 99–100.

6 Meyer, *Return of the God Hypothesis*, 234–36.

7 Ibid., 236.

8 John Leslie, "Anthropic Principle, World Ensemble, Design," *American Philosophical Quarterly* 19, no 2 (April 1982):141–51, at 150.

Notes

Chapter 5: Challenges to Materialism

1. William James, "Review of 'Human Personality and Its Survival of Bodily Death,'" *Proceedings of the Society for Psychical Research* 18, no. 46 (June 1903), reprinted in Gardner Murphy and Robert Ballou, eds., *William James on Psychical Research* (August M. Kelley, 1973), 225–239, at 227.
2. Foreword by Freeman Dyson in Elizabeth Lloyd Mayer, *Extraordinary Knowing: Science, Skepticism, and the Inexplicable Powers of the Human Mind* (Bantam, 2007), 2
3. Edmund Gurney, Frederic W. H. Myers, and Frank Podmore, *Phantasms of the Living*, vols. 1 and 2 (London: Rooms of the Society for Psychical Research, 1886). It is downloadable on Kindle for a charge and free at the Internet Archive (archive.org).
4. That statement also applies to other capabilities that are sometimes called "senses": proprioception, equilibrioception, thermoception, nociception, chronoception, and interoception.
5. Some of the most bizarre NDE cases involve an encounter with a person in the afterlife whom the experiencer had not known was dead. See Bruce Greyson *After: A Doctor Explores What Near-Death Experiences Reveal About Life and Beyond* (St. Martin's Essentials, 2021), 131–36.
6. The prospective studies include P. van Lommel et al., "Near-Death Experiences in Survivors of Cardiac Arrest: A Prospective Study in the Netherlands," *Lancet* 358, no. 9298 (December 15, 2001):2039–45; Janet

Schwaninger et al., "A Prospective Analysis of Near-Death Experiences in Cardiac Arrest Patients," *Journal of Near-Death Studies* 20 (June 2002):215–32; S. Parnia et al., "A Qualitative and Quantitative Study of the Incidence, Features and Aetiology of Near Death Experiences in Cardiac Arrest Survivors," *Resuscitation* 48 (February 2001):149–56; Bruce Greyson, "Incidence and Correlates of Near-Death Experiences in a Cardiac Care Unit," *General Hospital Psychiatry* 25 (July–August 2003):269–76; and P. Sartori, "A Long-Term Prospective Study to Investigate the Incidence and Phenomenology of Near-Death Experiences," *Network Review* 90 (spring 2006):23–25.

7   Alexander Batthyány, *Threshold: Terminal Lucidity and the Border of Life and Death* (St. Martin's Essentials, 2023), 105–6.

8   Michael Nahm and Bruce Greyson, "Terminal Lucidity in Patients with Chronic Schizophrenia and Dementia: A Survey of the Literature," *The Journal of Nervous and Mental Disease* 197, no. 12 (December 2009):942–44.

9   A collection of clips from the conference can be found at http://integral-options.blogspot.com/2012/01/nour-foundation-beyond-mind-body.html.

10 George A. Mashour et al., "Paradoxical Lucidity: A Potential Paradigm Shift for the Neurobiology and Treatment of Severe Dementias," *Alzheimer's and Dementia* 15 (August 2019):1107–14.

11 Another book on terminal lucidity, in German, is Michael Nahm's *Wenn die Dunkelheit ein Ende findet: Terminale Geistesklarheit und andere ungewöhnliche*

Phänomene in Todesnähe (Crotona Verlag, 2012).

12 The website for the Division of Perceptual Studies is https://med.virginia.edu/perceptual-studies.

13 William James, "What Psychical Research Has Accomplished," *The Will to Believe and Other Essays* (1897), reprinted in Gardner Murphy and Robert Ballou, eds., *William James on Psychical Research* (August M. Kelley, 1973), 25–47, at 41.

Chapter 6: A Strange New Respect

1 Charles Murray, *Human Accomplishment: The Pursuit of Excellence in the Arts and Sciences, 800 B.C. to 1950* (HarperCollins, 2003).

2 Readers on the political right may recognize the allusion in the title of this chapter. In the 1980s and 1990s, conservative public figures who moved left were likely to win "a strange new respect" from liberal journalists for their more enlightened behavior (or sometimes they were said to have "grown in office"). The first use of the phrase that I have been able to find is Tom Bethell's column, "Capital Ideas," in *The American Spectator* for January 1981. Bethell subsequently instituted annual "Strange New Respect Awards."

3 The book was Charles Murray, *In Pursuit: Of Happiness and Good Government* (Simon & Schuster, 1988).

4 John Rawls, *A Theory of Justice* (Belknap Press, 1971), 426.

5 Rodney Stark, *For the Glory of God: How Monotheism*

Notes

      Led to Reformations, Science, Witch-Hunts, and the End of Slavery (Princeton University Press, 2003).

6   Murray, *Human Accomplishment*, 618.

7   Stark, *For the Glory of God*, 147.

8   Alfred North Whitehead, *Science and the Modern World* (Free Press, 1967), 13. Quoted in Stark, *For the Glory of God*, 148.

9   Stark, *For the Glory of God*, 123. At this point you may wonder if Stark himself was a polemicist. As far as his own religious beliefs are concerned, Stark states in his introduction to *For the Glory of God* that "because this is a work of social science, not philosophy, I have taken pains neither to imply nor to deny the existence of God. This is a matter beyond the scope of science. Consequently, my personal religious beliefs are of concern only to me." In one of his few published statements of his religious views, he told an interviewer that "through most of my career, however, including when I wrote *The Rise of Christianity*, I was an admirer, but not a believer. I was never an atheist, but I probably could have been best described as an agnostic." He reports that he later felt able to acccpt an appointment at Baylor, the world's largest Baptist University, because by that time [2004] he had evolved into an "independent Christian." ("A Christmas Conversation with Rodney Stark," Center for Studies on New Religions, https://www.cesnur.org/2007/mi_stark.htm).

10 Jacques Barzun, *From Dawn to Decadence: 1500 to the Present: 500 Years of Western Cultural Life* (Harpers, 2000).

11   Percy Bysshe Shelley, *A Defence of Poetry* (1821).
12   Quoted in Murray, *Human Accomplishment*, 424.
13   Ibid., 457.
14   Endnote 35 for chapter 19 (page 619) in *Human Accomplishment* describes how I saw myself as of 2003: "Since I am arguing that the Christian religion is a primary force behind modern human accomplishment, readers may reasonably ask whether I am writing out of personal religious conviction. The answer is that I was raised as a mainstream Presbyterian, was drawn to Buddhism during the six years I lived in Asia (and still am), currently attend Quaker meetings, and can best be described as an agnostic."

Chapter 7: Enter C. S. Lewis

1   C. S. Lewis, *Mere Christianity* (Geoffrey Bles, 1952), adapted from 1941 radio talks and originally published in three volumes beginning in 1942. Quotations are taken from the 2001 HarperSanFrancisco edition; see the full citation in chapter 8, note 1, below.

Chapter 8: The Moral Law

1   C. S. Lewis, *Mere Christianity* (HarperSanFrancisco, 2001), 4.
2   Ibid., 6.
3   Lewis, *Mere Christianity*, 10–11.
4   W. D. Hamilton, "The Genetical Evolution of Social Behaviour," *Journal of Theoretical Biology* 7, no. 1 (July

Notes

1964):1–16.

5 Robert L. Trivers, "The Evolution of Reciprocal Altruism," *The Quarterly Review of Biology* 46, no. 1 (March 1971):35–57.

6 Francis S. Collins, *The Language of God: A Scientist Presents Evidence for Belief* (Free Press, 2006), 27. I read *The Language of God* soon after it was published in 2006 but had not picked it up for eighteen years when I began writing this book. Rereading Collins's opening chapters now, I'm a little embarrassed at how closely the narrative of my early steps tracks with his. It's not intentional. My best guess is that our ground-zero states of mind weren't that different.

7 The incident had a curious aftermath that in retrospect may support the Moral Law. When the car's driver was lying on the grass, she regained consciousness and began to choke. I couldn't make myself bend down and stick my finger down her throat to try to clear the passageway. At the time, I saw it as a failure on my part—my momentary courage had run out. But, looking back, the inner voice was no longer so imperious, and perhaps for a good reason. By the time she had regained consciousness, other people had gathered around her. It's not clear that sticking my finger down her throat was medically the right thing to do. It was probably best for the victim's sake that I let someone else, probably better qualified, take over. I didn't think it through at the time, but perhaps the reason that the inner voice was no longer so imperious was that my failure to intervene in the second

instance violated my male vanity, not the Moral Law. This logic is obviously speculative, but I can't get it out of my head.

8   Lewis, *Mere Christianity*, 24.
9   This and all quotations of the Bible, except where otherwise noted, are from The New Oxford Annotated Bible: New Revised Standard Version with the Apocrypha, ed. Michael J. Coogan (Oxford University Press, 2018).

Chapter 9: Who Wrote the Gospels and When?

1   C. S. Lewis, *Mere Christianity* (HarperSanFrancisco, 2001), 52.
2   Bart D. Ehrman, *The New Testament: A Historical Introduction to the Early Christian Writings* (Oxford University Press, 1997), 72–73.
3   Albert Schweitzer, *The Quest of the Historical Jesus: From Reimarus to Wrede*, trans. William Montgomery (A. and C. Black, 1910), originally published in German in 1906 under the title *Von Reimarus zu Wrede: eine Geschichte der Leben-Jesu-Forschung*.
4   This lost text, which is known as *The Diatessaron* ("Through Four"), harmonized the four canonical Gospels into a single narrative. Peter J. Williams, *Can We Trust the Gospels?* (Crossway, 2018), 38.
5   Quoted in Richard Bauckham, *Jesus and the Eyewitnesses: The Gospels as Eyewitness Testimony* (Eerdmans, 2006), 15–16.
6   Quoted in ibid., 35.

Notes

7   Quoted in Michael F. Bird, *The Gospel of the Lord: How the Early Church Wrote the Story of Jesus* (Wm. B. Eerdmans Publishing Co., 2014), 215. The quotation is taken from an extract included in Eusebius, *Ecclesiastical History*. The original source is among Clement's works that have been lost—a useful reminder of how much more material the patristic writers had to work with than modern scholars have.

8   Literacy became common among Jews after the Temple was destroyed in 70 CE and Judaism transitioned from being a religion of the Temple to a religion of the Book. In the apostles' generation, literacy was rare. The only apostle other than Matthew who might have been literate was John, whose family had servants and therefore might have been wealthy enough to give John some formal education.

9   Quoted in Bauckham, *Jesus and the Eyewitnesses*, 203. *Logia* is a Greek word translated variously as "sayings," "oracles," or "discourses." The English word *oracles* refers to divine revelations communicated to humans.

10   Quoted in Bird, *The Gospel of the Lord*, 214.

11   Quoted in ibid., 215.

12   Quoted in ibid., 214.

13   Quoted in Bernard Orchard and Harold Riley, *The Order of the Synoptics: Why Three Synoptic Gospels?* (Mercer University Press, 1987), 125.

14   Quoted in Bird, *The Gospel of the Lord*, 213.

15   Quoted in Bauckham, *Jesus and the Eyewitnesses*, 434.

16   Quoted in Bird, *The Gospel of the Lord*, 214. At this point in John's narrative of the Last Supper, "the

disciple Jesus loved" asked Jesus who would betray him. In the King James version, "He then lying on Jesus' breast saith unto him, 'Lord, who is it?'" (John 13:25).

17   Quoted in Bird, *The Gospel of the Lord*, 214.
18   Jonathan Bernier, *Rethinking the Dates of the New Testament: The Evidence for Early Composition* (Baker Academic, 2022), 35, 87.
19   John A. T. Robinson, *Redating the New Testament* (1976), 336. Emphasis in the original.
20   Quoted in ibid., 343.
21   Ibid., 85.
22   Ibid., 81.
23   Bernier, *Rethinking the Dates of the New Testament*, 11.
24   Ibid., 275.
25   Adolf von Harnack, *The Date of the Acts and of the Synoptic Gospels* (Williams & Norgate and G.P. Putnam's Sons, 1911), quoted in Bernier, *Rethinking the Dates of the Gospels*, 62.

Chapter 10: The Historicity of the Gospels

1   I was once mugged in New York City. The police caught the perpetrator (I can be confident of that because the suspect, picked up within half an hour of my report, had distinctive items from my wallet in his possession; for example, a 20-baht Thai banknote). I had made two mistakes when I told the police about it. The perpetrator had whispered during the robbery, and I thought he had a Spanish accent. I was wrong. I also thought that the perpetrator had been holding a

Notes

knife to my back. It was a handgun.

2 The information on geography is just a portion of the fact-checking material presented in Peter J. Williams, *Can We Trust the Gospels?* (Crossway, 2018), 52–63.
3 Ibid., 67.
4 Ibid., 68–69.
5 Ibid., 69.
6 This definition is adapted from Lydia McGrew, *Hidden in Plain View: Undesigned Coincidences in the Gospels and Acts* (Deward Publishing, 2017), 18.
7 Richard Bauckham, *Jesus and the Eyewitnesses: The Gospels as Eyewitness Testimony* (Eerdmans, 2006), chapter 10.
8 Williams, *Can We Trust the Gospels?*, chapter 6. In his analysis of the differences that do exist between Erasmus's version of the New Testament Greek text and the Greek text used in the twenty-first century, Williams points out that Erasmus himself was aware of many of the verses that were problematic.
9 If height were perfectly normally distributed, the figure would be one out of every 31,574 people, but height, like many normally distributed variables, has "fat tails," meaning that the extremes occur somewhat more frequently than the mathematics of a normal distribution predicts.
10 Arthur C. Clarke's Third Law, published in several of Arthur C. Clarke's works: "Any sufficiently advanced technology is indistinguishable from magic."
11 The seven undisputed letters are Romans, 1 and 2 Corinthians, Galatians, Philippians, 1 Thessalonians, and Philemon. The disputed letters are Ephesians,

Colossians, and 2 Thessalonians. The three "pseudepigraphical" letters are Titus and 1 and 2 Timothy. This classification is based on the "Authorship" section for each of the Pauline letters in The New Oxford Annotated Bible: New Revised Standard Version with the Apocrypha, ed. Michael D. Coogan (Oxford University Press, 2018). Here is the biographical timeline that Paul describes in Galatians 1:13–24, 2:1–2. "You have heard, no doubt, of my earlier life in Judaism. I was violently persecuting the church of God and was trying to destroy it. I advanced in Judaism beyond many among my people of the same age, for I was far more zealous for the traditions of my ancestors. But when God, who had set me apart before I was born and called me through his grace, was pleased to reveal his Son to me, so that I might proclaim him among the Gentiles, I did not confer with any human being, nor did I go up to Jerusalem to those who were already apostles before me, but I went away at once into Arabia, and afterwards I returned to Damascus. Then after three years I did go up to Jerusalem to visit Cephas and stayed with him fifteen days; but I did not see any other apostle except James the Lord's brother. In what I am writing to you, before God, I do not lie! Then I went into the regions of Syria and Cilicia, and I was still unknown by sight to the churches of Judea that are in Christ; they only heard it said, 'The one who formerly was persecuting us is now proclaiming the faith he once tried to destroy.' And they glorified God because of me. Then after fourteen years I went

Notes

up again to Jerusalem with Barnabas, taking Titus along with me. I went up in response to a revelation. Then I laid before them (though only in a private meeting with the acknowledged leaders) the gospel that I proclaim among the Gentiles, in order to make sure that I was not running, or had not run, in vain."

12   See Bauckham, *Jesus and the Eyewitnesses*, 264–71.
13   Ibid., 266.
14   C. H. Dodd, quoted in ibid.
15   It is important to this logic that many witnesses be available and that the encounters involve either Jesus's physical presence or a prolonged vision and interaction of the kind that Paul experienced. In *How Jesus Became God*, Bart Ehrman suggests that Peter, Mary Magdalene, and perhaps a few others had hallucinatory visions, but a few people (one of them a woman, and therefore to be discounted) reporting a vision doesn't seem sufficient to explain all the disciples' continuing insistence that Jesus had risen from the dead, or to explain Christianity's electrifying appeal.
16   Bauckham, *Jesus and the Eyewitnesses*, 39.
17   N. T. Wright, *The Resurrection of the Son of God* (Fortress Press, 2003), Part I.
18   My description of what an artist would have to do to create the image on the Shroud of Turin by hand is adapted from Giulio Fanti and Pierandrea Malfi, *The Shroud of Turin: First Century After Christ*, 2nd ed. (Jenny Stanford Publishing, 2020), 26–27. Here is their more elaborate version: "First of all, the artist should dip the brush, not in the color, because there are not

pigments on the threads, but in an acid capable of shading the linen chemically.... Since colored fibers are side-by-side with uncolored ones, the brush must have only one bristle with a diameter not superior to 0.01 mm (0.00039 in.). Inexplicably, the artist also has to be able to color the part of the straw on the inner side of the bundle without coloring the adjacent straws, since the color is uniformly distributed around the circumference. Then the acid has to be placed on the fiber just for a split second because it must have no time to act in depth and undermine the uncolored cellulose. Finally, the artist would have to paint in the same way all the million straw-fibers that constitute the shroud using a microscope (not existing in the Middle Ages)."

19 L. A. Schwalbe and R. N. Rogers, "Physics and Chemistry of the Shroud of Turin: A Summary of the 1978 Investigation," *Analytica Chimica Acta* 135, no. 1 (January 1982):3–49, at 45, available at shroud.com.

20 Raymond N. Rogers, "Studies on the Radiocarbon Sample from the Shroud of Turin," *Thermochimica Acta* 425, nos. 1–2 (January 2005):189–94, at 193, available at shroud.com. The choice of the sample, described in detail by Fanti and Malfi in *The Shroud of Turin*, was badly botched. Apart from Rogers's analyses, it was discovered that the fabric used for the dating had been attached to the shroud to repair areas of its cloth that had been worn down over the centuries, and that this new addition to the shroud had been dyed with a dye that was not available in Europe until 1291.

Notes

21 Liberato De Caro et al., "X-ray Dating of a Turin Shroud's Linen Sample," *Heritage* 5, no. 2 (April 2022):860–70, at 80, available at shroud.com.
22 John Polkinghorne, *The Faith of a Physicist: Reflections of a Bottom-Up Thinker* (Princeton University Press, 1994), 122.

Chapter 11: What's the Point?

1 Richard P. Feynman, *The Character of Physical Law* (MIT Press, 1965). The original statement of this point was in the 1964 Messenger Lectures at Cornell.
2 N. David Mermin, "What's Wrong with This Pillow?" *Physics Today* 42, no. 4 (April 1989):9–11, at 9.
3 "Paradigm shift" is Thomas Kuhn's famous label for cycles of scientific progress. In any era, a reigning paradigm defines the framework within which normal scientific advances continue. The first scientists who report anomalies are likely to be ignored or chastised. As time goes on, new anomalies inconsistent with the paradigm come to light. Eventually, Kuhn argues, the result is a paradigm shift to a new received wisdom. Thomas S. Kuhn, *The Structure of Scientific Revolutions* (University of Chicago Press, 2012). This fiftieth-anniversary edition includes Kuhn's response to his critics and an introductory essay by Ian Hacking.
4 C. S. Lewis, *The Abolition of Man* (Oxford University Press, 1943), 9.
5 During the same period, in Greek Ionia, Heraclitus introduced the concept of the *logos*, a cosmic principle

Notes

of reason and order that would subsequently become a central aspect of Christian thinking. But Heraclitus's treatment of the *logos* had nothing to do with religion or ethics.

6   Richard Dawkins, *River Out of Eden: A Darwinian View of Life* (Basic, 1995), 133.

7   My published libertarian positions may lead some to wonder about my endorsement of the central role of *agape* and, by extension altruism (a suspiciously bleeding-heart concept in many libertarian and objectivist circles). I see no contradiction between believing in a Moral Law that commands me to be altruistic and opposing man-made laws that farm out my obligation to government programs.

8   Pim van Lommel, *Consciousness Beyond Life: The Science of the Near-Death Experience* (HarperOne, 2010), 29–30. Van Lommel puts the proportion of those who have a "hell" experience at 1–2 percent.

9   John Polkinghorne, *The Faith of a Physicist: Reflections of a Bottom-Up Thinker* (Princeton University Press, 1994), 130.

Index

Abraham, 26
afterlife, 45, 55, 153–55, 164n5
agape, 81, 153, 178n7
agnosticism, 1, 2, 3, 7, 8, 32, 63, 145, 167n9
Albertus Magnus, 68
altruism, 81, 82–83, 149, 178n7
anthropic principle, 36
anthropomorphization, 26–27, 156. *See also* deanthropomorphization
Apollo program, 10
Apostles' Creed, 8, 9
Aquinas, Thomas, 2, 27, 65, 67, 68
Arabia, 64
Aristotle, 25, 27, 64
arts, the, 69–72
atheism, 16, 68, 145, 167n9
Athens, Greece, 64
autonomy, 17, 66

Bach, Johann Sebastian, 72
Banfield, Edward, 9
Barzun, Jacques, 69
Batthyány, Alexander, 54–55
Bauckham, Richard, 90–91, 92, 93, 115–16, 119, 129, 130
Baur, Ferdinand Christian, 100
Beatles, The, 15
Beethoven, Ludwig van, 15, 70
Bennett, Bill, 10, 73
Berkeley, George, 43
Bernier, Jonathan, 106–7
Big Bang, 29–44, 63, 144
black holes, 39, 41
Boyle's Law, 22
Brooks, David, 7
Bruno, Giordano, 68
Buddha, the, 148, 158
Buddhism, 15, 148–51, 152, 168n14

179

Index

Buridan, Jean, 68
Burkittsville, Maryland, 10, 11
Bush, George W., 73

Capernaum, 112, 114
Carter, Brandon, 36
Catholicism, 1, 3, 31, 66, 67–68
China, 2, 64, 78, 148
civilization, 64, 67, 78
Clarke, Arthur C., 125, 173n10
Clement of Alexandria, 94, 96, 97, 98, 100, 171n7
coincidences, undesigned, 117–19
Collins, Francis, 35, 36, 81, 83, 84, 122, 169n6
Columbus, Christopher, 68
communism, 82
consciousness, 21, 30, 45–46, 50, 51, 53, 55, 57, 58, 59, 63, 145–48, 154, 169
conservatism, 1, 166n2
Constantinople, 133, 137
Copernicus, Nicolaus, 68
creation, 25, 27, 29, 31, 32–33, 146

dark matter, 39
Darwin, Charles, 142
Dawkins, Richard, 68, 152
de-anthropomorphization, 26–28
dementia, 54–55, 56
Derrida, Jacques, 91
Dirac, Paul, 22
Duchamp, Marcel, 71
Dyson, Freeman, 48, 50

Edessa, Turkey, 133, 137
Ehrman, Bart, 87, 175n15
Einstein, Albert, 30, 145
electromagnetic force, 34
Enlightenment, the, 45, 47, 103
Erasmus, Desiderius, 121, 173n8
ethics, 148–50, 177n5
Eusebius, 94, 96
Evangelicalism, 73, 122
evolution, theory of, 7, 52, 66, 80–83, 150
extrasensory perception (ESP), 48
eyewitness accounts, the Gospels as, 90–93, 95, 99, 102, 111, 117, 119, 123, 128, 129, 130, 140

Index

Farrer, Austin, 102
Feynman, Richard, 143–44
fission, nuclear, 35
Florence, Italy, 64
form criticism, 88–89
Foucault, Michel, 91
Freud, Sigmund, 142
fusion, nuclear, 37

Galilei, Galileo, 68
Galileo's Law, 22
general relativity, 30, 145, 146
Gerson, Michael, 7
"God of the Gaps", 141–48
Gospels, the: authenticity of, 111–40; authorship of, 85–110; dating of, 100–110; geography in, 113–15; Jewish names in, 115–17; Jewishness of, 105–6; John, 86, 93, 97, 99–100, 101, 102, 105, 107, 114, 125, 131; Luke, 86, 93, 99, 101, 102, 107, 114; Mark, 86, 93, 98–99, 101, 102, 107, 114; Matthew, 86, 93, 97–98, 101, 102, 107, 114
Gospels, apocryphal, 115, 117
gravity, 34

Harnack, Adolf von, 100–101, 109
Harvard University, 2, 67
Hawking, Stephen, 24
hell, 154, 178n8
Heraclitus, 177n5
Herrnstein, Richard, 11, 13
Hilbert, David, 22
Hilo, Hawaii, 14
Hinduism, 78, 148–51, 152
Hobbes, Thomas, 68
Homer, 63
Hoyle, Fred, 32
Hubble, Edwin, 31
Hubble's Law, 31
Huxley, T. H., 8

Ilan, Tal, 115
individualism, 64–65
interconnectedness, 152
IQ, 11, 16, 30, 161n1
Irenaeus, 94, 95–97, 98, 99

James, William, 46–47, 49, 58, 146
Jesus Christ: birth of, 123; divinity of, 86, 122, 129, 156; miracles of, 121–25; resurrection of, 125–33; teachings of, 119–21

Index

Job, 26
John the Elder, 98
Johnson, Samuel, 43–44
Josephus, 115
Judaism, 45, 65, 105–6, 149–51, 171n8, 174n11
Justin Martyr, 94, 95, 98

Krauthammer, Charles, 23, 73, 162n2

Laozi, 148–49, 158
Laplace, Pierre-Simon, 142
Lemaître, Georges, 31
Leslie, John, 43
Lewis, C. S., 73–75, 77–80, 83–84, 85–86, 122, 141, 148, 152, 156
libertarianism, 1, 178n7
life after death. *See* afterlife
Lightfoot, J. B., 100
Lincoln, Abraham, 11
Lincoln, Virginia, 11
Lirey, France, 133, 137
logos, 177n5
lucidity, terminal, 50, 53, 54–57, 59, 145–47
Lyell, Charles, 142

Maharishi Mahesh Yogi, 15
Maryland, 9
materialism: 33, 145, 147; challenges to, 45–59
mathematics, 18–19, 22–23, 37, 41, 143, 156
McConnell, Lowell, 157
meditation, 14–15, 17, 152
Mermin, David, 144
metaphysics, 23
Methodism, 8
Meyer, Stephen, 33, 36, 41
Michelangelo, 26
Michelson-Morley experiment, 145
miracles, 121–25
modernity, Western, 17
Moody, Raymond, 50, 52, 53
moral law, 77–84, 85, 141, 148–51
multiverse theory, 41–42, 44
music, 15–16, 69, 71, 72, 161n1
mystery, 24, 25, 122

names, disambiguation of, 116–17
Napoleon Bonaparte, 142
near-death experiences (NDEs), 50–54, 57, 59, 145, 147, 152, 153–54, 164n4

Index

Nebuchadnezzar, 104
Nero, 102–3, 107
neuroscience, 146
Newton, Iowa, 8, 13, 157
Newton, Isaac, 88, 143, 145–46
Newton's Second Law of Motion, 22
Novak, Michael, 3
nuclear force, strong, 34
nuclear force, weak, 34

Ohm's Law, 22
Ockham, William of, 68
Oresme, Nicole, 68
Origen, 96

Papias, 94–96, 97, 98
paranormal phenomena, 45–46, 47, 49, 50, 57, 58, 146
particles, 32, 33, 39–40, 41, 143
Pascal's Wager, 155
patristics, 90–93, 96–100, 171n7
Paul, 97, 98, 99, 107–8, 109, 128, 129, 174, 175n15
Penrose, Roger, 40–41, 42, 43
Penrose-Hawking singularity theorems, 40
Penzias, Arno, 32

Peter, 95, 97, 98–99, 107–8, 109, 112–13, 116, 125, 128, 130, 132
Peter the Monk, 15
Planck's Law, 143
Polkinghorne, John, 74, 140, 156
Polycarp, 95–96
postmodernism, 69, 89
pragmatism, 46
Presbyterianism, 2, 8, 157, 168n14
Protestantism, 9, 66, 73
psi phenomena. *See* paranormal phenomena
Psychical Research, Society for (SPR), 47, 49
psychology: 52, 83, 109; evolutionary, 80, 150
Pythagoreanism, 95

"Q" (hypothesised source for the Gospels), 108, 123
Quakerism, 8–9, 10–11, 14, 27, 168n14
quantum mechanics, 22, 41, 143–44, 146

Rawls, John, 64
Reagan, Ronald, 73, 101

Index

redaction criticism, 89, 106, 119, 140
Rees, Martin, 36–41, 42
reincarnation, 57, 147
repentance, 153
revelation, 3
revisionism, Biblical, 87–90, 91–92, 94, 99, 109, 129, 140
Rhine, Joseph, 48
Robinson, John, 102, 103, 104–5
Rogers, Raymond, 138, 176n20
Rome, Italy, 95, 97, 106, 107, 108, 109, 113, 116, 126
Russell, Bertrand, 67

Sagan, Carl, 49, 68
salvation, 65, 147
Schoenberg, Arnold, 71
scholasticism, 67
Schweitzer, Albert, 88
science: 8, 21, 29, 31, 32, 42, 46–47, 48, 49, 53, 72, 121, 135, 139, 141–43, 144, 155; and Christianity, 66–69, 74
Scientific Revolution, the, 68
Second Temple, destruction of, 103–5
secularism, 1, 2, 11, 21, 25, 70, 72, 78, 88, 138
Serrano, Andres, 71
Shakespeare, William, 16
Shroud of Turin, 133–40
Shroud of Turin Research Project (STURP), 134–39
sin, 142, 147, 153
social media, 55
soul, 45, 46, 59, 97, 147, 148, 155
spirituality, 16–17, 161n1
Stark, Rodney, 66–69
stars, formation of, 35, 37, 38, 39, 40
Stoicism, 95
suicide, 155

Taoism, 148–51, 152
Tel Aviv, Israel, 111
Thailand, 14–15
Thomism, 66
time, 27, 32, 33
Titus, 104
transubstantiation, 3
Turin, Italy, 133

Unitarianism, 9
universities, medieval, 67–68

Warhol, Andy, 71

Index

Washington, DC, 8, 10, 82
Wehner, Pete, 73
Whitehead, Alfred North, 67
Williams, Peter, 93, 106, 114,
 116, 121, 122
Wilson, Robert, 32
Wright,n T., 129, 131